Lecture Notes of the Institute for Computer Sciences, Social Informatics and Telecommunications Engineering 375

More information about this series at http://www.springer.com/series/8197

João L. Afonso · Vitor Monteiro ·
José Gabriel Pinto (Eds.)

Sustainable Energy
for Smart Cities

Second EAI International Conference, SESC 2020
Viana do Castelo, Portugal, December 4, 2020
Proceedings

 Springer

Editors
João L. Afonso (iD)
Department of Industrial Electronics
University of Minho
Guimaraes, Portugal

Vitor Monteiro (iD)
Department of Industrial Electronics
University of Minho
Guimaraes, Portugal

José Gabriel Pinto (iD)
University of Minho
Guimaraes, Portugal

ISSN 1867-8211 ISSN 1867-822X (electronic)
Lecture Notes of the Institute for Computer Sciences, Social Informatics
and Telecommunications Engineering
ISBN 978-3-030-73584-5 ISBN 978-3-030-73585-2 (eBook)
https://doi.org/10.1007/978-3-030-73585-2

This Springer imprint is published by the registered company Springer Nature Switzerland AG
The registered company address is: Gewerbestrasse 11, 6330 Cham, Switzerland

Preface

We are honored to present the proceedings of the second edition of the International Conference on Sustainable Energy for Smart Cities (SESC), sponsored by the European Alliance for Innovation (EAI) in collaboration with the University of Minho, Portugal. This second edition of the SESC conference was framed within the sixth annual Smart City 360° Viana do Castelo, Portugal, December 4, 2020. The main objective of the SESC 2020 conference was to provide a multidisciplinary scientific meeting toward answering the complex technological problems of emergent smart cities. The subjects related to sustainable energy, framed with the scope of smart cities and addressed at the SESC 2020 conference, are crucial to obtaining an equilibrium between economic growth and environmental sustainability, as well as to reducing the impact of climate change.

The SESC 2020 technical program contained 13 full papers along with the main conference tracks, which were organized into 4 presentation sessions. For each submitted paper, a double-blind peer review process, with a minimum of three reviews, was followed. We sincerely thank all the members of the Technical Program Committee for assistance in the peer review process, contributing to a high-quality technical program. We would also like to thank all the external reviewers from numerous countries around the world, who were specially selected according to the several areas of expertise covered in the SESC 2020 conference. Last but not least, we are grateful to the other members of the Steering and Organizing Committees, and to the EAI, whose collaboration was fundamental for the success of the conference.

The SESC 2020 conference was a notable scientific meeting for all researchers, developers, and practitioners, offering the opportunity to discuss all scientific and technological directions targeting the smart city paradigm. As a consequence of the success of the SESC 2020 conference and as evidenced by the papers presented in this volume, covering distinct areas of research, we are looking forward to a successful and stimulating future series of SESC conferences.

João L. Afonso
Vítor Monteiro
José Gabriel Pinto

Conference Organization

Steering Committee

Imrich Chlamtac	University of Trento, Italy
João L. Afonso	University of Minho, Portugal
Vítor Monteiro	University of Minho, Portugal
Gabriel Pinto	University of Minho, Portugal

Organizing Committee

General Chair

João L. Afonso   University of Minho, Portugal

General Co-chairs

Vítor Monteiro	University of Minho, Portugal
Gabriel Pinto	University of Minho, Portugal

Technical Program Committee Chair

Carlos Couto   University of Minho, Portugal

Sponsorship and Exhibit Chair

Paula Ferreira   University of Minho, Portugal

Local Chair

Madalena Araújo   University of Minho, Portugal

Workshops Chair

Delfim Pedrosa   University of Minho, Portugal

Publicity and Social Media Chair

Luís Barros   University of Minho, Portugal

Publications Chair

Tiago Sousa   University of Minho, Portugal

Web Chair

Mohamed Tanta   University of Minho, Portugal

Technical Program Committee

Edson H. Watanabe	Federal University of Rio de Janeiro, Brazil
João A. Peças Lopes	University of Porto, Portugal
Rune Hylsberg Jacobsen	Aarhus University, Denmark
Richard Stephan	Federal University of Rio de Janeiro, Brazil
Carlos Hengeler	Antunes University of Coimbra, Portugal
Adriano Carvalho	University of Porto, Portugal
Marcelo Cavalcanti	Federal University of Pernambuco, Brazil
João P. S. Catalão	University of Porto, Portugal
Jose A. Afonso	University of Minho, Portugal
Antonio Lima	State University of Rio de Janeiro, Brazil
António Pina Martins	University of Porto, Portugal
Hfaiedh Mechergui	University of Tunis, Tunisia
Stanimir Valtchev	NOVA University Lisbon, Portugal
Rosaldo Rossetti	University of Porto, Portugal
Chunhua Liu	City University of Hong Kong, China
Luis Monteiro	University of Rio de Janeiro, Brazil
João P. P. Carmo	University of São Paulo, Brazil
A. Caetano Monteiro	University of Minho, Portugal
Marcello Mezaroba	Santa Catarina State University, Brazil
João Martins	NOVA University Lisbon, Portugal
J. Aparicio Fernandes	University of Minho, Portugal
Paulo Pereirinha	Polytechnic Institute of Coimbra, Portugal
Jelena Loncarski	Polytechnic University of Bari, Italy
Amira Haddouk	University of Tunis, Tunisia
Joao C. Ferreira	ISCTE – University Institute of Lisbon, Portugal
Stefani Freitas	Federal University of Tocantins, Brazil
Julio S. Martins	University of Minho, Portugal
L. Pecorelli Peres	State University of Rio de Janeiro, Brazil
Kleber Oliveira	Federal University of Paraíba, Brazil
M. J. Sepúlveda	University of Minho, Portugal
Orlando Soares	Polytechnic Institute of Bragança, Portugal
José L. Lima	Polytechnic Institute of Bragança, Portugal
Carlos Felgueiras	Polytechnic of Porto - School of Engineering (ISEP)

Contents

Power Electronics

Design of Current Controllers for Three Phase Voltage PWM Converters for Different Modulation Methods

Rodrigo Guzman Iturra$^{(\boxtimes)}$ (iD) and Peter Thiemann

South Westphalia University of Applied Sciences,
Lübecker Ring 2, 59494 Soest, Germany
{guzmaniturra.rodrigozenon,thiemann.peter}@fh-swf.de
https://www4.fh-swf.de/de/home/

Abstract. Grid Tie Three Phase Voltage PWM converters can be conceived as current sources that inject currents into the grid at the point of common coupling (PCC). In order to achieve a good performance, the voltage source inverter (VSI) should be commanded by a current controller to track as accurate as possible a current reference. Pulse with modulation (PWM) and space vector modulation (SVM) are two techniques used in VSIs to generate the time average output voltage demanded by the current controller. By using the average model of the VSI, controlled either by PWM or by SVM, we can use feedback linear control for the analysis and design of the current controller. The main goal of this technical paper is to illustrate an analytical formula to calculate the gains of the proportional-integral current controller. The gains are calculated based on the values of the coupling inductivity, the DC Bus voltage and the switching frequency and given for two PWM modulation methods.

Keywords: Current controller · PWM modulation · Space vector modulation · Voltage source inverter

1 Introduction

There is large variety of applications that rely on DC/AC converters such as ac motor drives, STATCOMs, active power filters, uninterruptible power supplies and photovoltaic systems [1,2]. The preferred topology for the DC/AC converters are voltage source inverters (VSIs). Even due converters based on the current source inverter topology (CSI) exist, the switching characteristics of IGBTs and power MOSFETs favore the VSI topology in terms of efficiency in comparison with CSI and made the VSI the predominant topology for DC/AC converters [3]. In many of former mentioned applications, the three phase currents that are generated by the DC/AC converter follow a reference provided by an outer loop. The current references are generated by speed controllers in the case of ac motor drives, a reactive power controller or grid voltage regulator in the

© ICST Institute for Computer Sciences, Social Informatics and Telecommunications Engineering 2021
Published by Springer Nature Switzerland AG 2021. All Rights Reserved
J. L. Afonso et al. (Eds.): SESC 2020, LNICST 375, pp. 3–14, 2021.
https://doi.org/10.1007/978-3-030-73585-2_1

case of STATCOMs and in case of photovoltaic systems they follow a reference generated based on the amount of active power that needs to be injected into the grid. Most of the DC/AC converters are based on a voltage source inverter (VSI) and in order to achieve a good performance, the VSI is commanded by a current controller to track as accurate as possible the three phase currents reference. Pulse with modulation (PWM) and space vector modulation (SVM) are two techniques used in VSIs to generate the time average output voltage demanded by the current controller. By using the average model of the VSI, controlled either by PWM or by SVM, we can use feedback linear control for the analysis and design of the current controller. The main goal of this document is to illustrate an analytical formula to calculate the gain for a proportional-integral current controller based on the values of the coupling inductivity, the DC Bus voltage and the switching frequency of a three phase PWM converter.

This paper is based on the work done by Holmes et al. in [4], however the former paper dealt just with the current controller design for two level converter with conventional PWM as modulation method. The small contribution of this paper is to extent the calculation of the current controller gains for two level converters with SVM modulation and for three level converters with PWM (e.g. Neutral Point Clamped NPC, T Type NPC and others). Moreover, this paper shows a more detailed mathematical derivation of the formulas step by step in constrast with the original on [4]. This paper is organized as follows: Sect. 2 describes the model of the three phase voltage PWM converter connected to the grid. Section 3 presents the derivation of the formula for the calculation of the proportional gain based on the parameters of the thee phase voltage PWM converter. Section 4 presents the experimental results obtained by setting up a prototype govern by a digital control system with the current controller proposed in this paper. Finally in Sect. 5, the conclusions are presented.

2   System Modeling

Figure 1 shows the structure of the DC/AC converter based on a VSI, which represents the power electronics part. Holmes et al. have shown in [4] that each phase can be controlled independently if the three phase system is balanced. Therefore considering each phase independently as an individual half-bridge, the dynamics of the AC current for the three phases can be described by [5]:

$$L_i \frac{di_{\mathrm{A}}(t)}{dt} + R_i i_{\mathrm{A}}(t) = v_{\mathrm{aI}}(t) - v_{\mathrm{GA}}(t) \tag{1}$$

$$L_i \frac{di_{\mathrm{B}}(t)}{dt} + R_i i_{\mathrm{B}}(t) = v_{\mathrm{bI}}(t) - v_{\mathrm{GB}}(t) \tag{2}$$

$$L_i \frac{di_{\mathrm{C}}(t)}{dt} + R_i i_{\mathrm{C}}(t) = v_{\mathrm{cI}}(t) - v_{\mathrm{GC}}(t) \tag{3}$$

Where v_{Gx} are the voltage of the grid and v_{xI} is the time average voltage produced by the inverter. Figure 2 shows the block diagram of the controlled system considering the time delay introduced by the modulation process. In $\alpha - \beta$ coordinates the model of the three phase converter is as follows:

$$L_i \frac{di_\alpha(t)}{dt} + R_i i_\alpha(t) = v_{\alpha I}(t) - v_{G\alpha}(t) \tag{4}$$

$$L_i \frac{di_\beta(t)}{dt} + R_i i_\beta(t) = v_{\beta I}(t) - v_{G\beta}(t) \tag{5}$$

The time average voltage produced by the inverter can be controlled directly by the duty cycle (m) feed to the PWM modulator. The duty cycle is the output of the current controller. The average model of the VSI is considered as a linear amplifier [5] with a gain G_{INV} related with the DC bus voltage v_{DC} and a dead time (transport delay) introduced by the switching process [6]. Thus:

$$v_{xI} = G_{INV} \cdot e^{-sT_d} \cdot m \tag{6}$$

Where for PWM with double sample update the delay can be approximated by $T_d = \frac{1}{2F_s}$, being F_s the switching frequency and $G_{INV} = \frac{V_{dc}}{2}$ for the case of PWM. For the controller design, many authors approximate the transport delay produced by the switching process by a first order Pade approximation [6]:

$$e^{-sT_d} \approx \frac{1 - s(T_d/4)}{1 + s(T_d/4)} \tag{7}$$

Or even they approximate the transport delay roughly by a first order system of the form [7]:

$$e^{-sT_d} \approx \frac{1}{1 + s(3 \cdot T_d/2)} \tag{8}$$

Equations (7) and (8) are a good approximation in case that the switching frequency is at least a factor of 10 times larger than the frequency of the reference signal feed to the PWM modulator. However, there is applications where this condition does not hold, for instance in high power applications when the switching frequency cannot be high due to unacceptable switching losses or in case of active power filters that compensate high order harmonics. In such cases the approximation will lead to a bad performance of the current controller or even to instabilities in the current control loop [8]. Therefore, for such applications is better to design the current controller considering the transport delay directly in the model.

2.1   Open Loop Transfer Function

By disregarding the voltage of the grid v_{Gx} that acts just a disturbance and can be almost entirely canceled by feedforward controller as is explained in [5], the open loop transfer function for the current loop considering the transport delay and a proportional controller is as follows:

$$G(s)_{OL} = K_P \cdot G_{INV} \cdot e^{-sT_d} \cdot \frac{1}{L_i s + R_i} \tag{9}$$

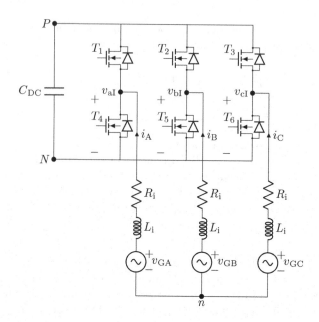

Fig. 1. Voltage Source Inverter (VSI) connected to a power grid through a coupling inductor L_i with an internal resistance R_i. The grid is represented by the three voltages sources $v_G x$.

The resistance of the coupling inductivity R_i that is usually smaller compared with the reactance can be neglected, it follows:

$$G(s)_{\mathrm{OL}} = K_{\mathrm{P}} \cdot G_{\mathrm{INV}} \cdot e^{-sT_{\mathrm{d}}} \cdot \frac{1}{L_i s} \tag{10}$$

3   Derivation of Proportional Gain

The following derivation follows the procedure described on [4], furthermore in this document the derivation is done in a step by step fashion and at the end a final analytical formula is achieved. The phase angle of the frequency dependent expression (10) is equal to:

$$\angle G(s)_{\mathrm{OL}} = \angle(K_{\mathrm{P}} \cdot G_{\mathrm{INV}}) + \angle(e^{-sT_{\mathrm{d}}}) + \angle(\frac{1}{sL_i}) \tag{11}$$

K_{P} and G_{INV} are just gains and their contribution to the total phase is 0°, the coupling inductivity is always contributing -90° or $-\pi/2$ rad to the total phase when R_i is neglected. The dead time is introducing negative phase in radians that is linearly proportional to the frequency [9], thus:

$$\angle G(s)_{\mathrm{OL}} = 0 - (\omega T_{\mathrm{d}}) - \frac{\pi}{2} \tag{12}$$

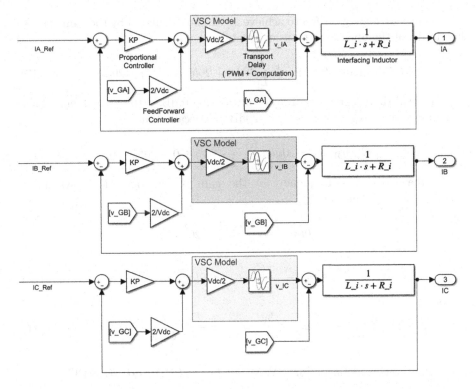

Fig. 2. Block diagram of the current control loop for the three phases. Feedforward and feedback controllers are shown.

The phase of the open loop transfer function can be made function of the desired phase margin (PM) according to:

$$\angle G(s)_{\text{OL}} = -\pi + PM \tag{13}$$

$$-\pi + PM = -(\omega T_{\text{d}}) - \frac{\pi}{2} \tag{14}$$

If we choose certain PM, we can use Eq. (14) to calculate which cross over frequency ω_c corresponds to the selected PM margin:

$$\omega_c = \frac{\frac{\pi}{2} - PM}{T_{\text{d}}} \tag{15}$$

A PM of 30° or $\pi/6$ rad, produces a close loop with a damping ratio of approximate 0.4 that should be adequate to track without that much error sinusoidal references and it is judged as the lowest adequate value for the phase margin according to [9]. For a PM = 30° and $T_{\text{d}} = \frac{1}{2F_{\text{s}}}$ in (15), it follows:

$$\omega_c = \frac{\pi}{3 \cdot T_{\text{d}}} = \frac{2 \cdot \pi \cdot F_{\text{s}}}{3} \tag{16}$$

The proportional gain K_P to achieve ω_c can be calculated by the gain expression in dB of the open loop transfer function:

$$| G(s)_{\text{OL}}|_{\text{dB}} = 20log(\frac{K_\text{P} \cdot G_{\text{INV}}}{L_\text{i}}) + 20log(e^{-sT_\text{d}}) + 20log(\frac{1}{s}) \qquad (17)$$

The dead time element is always contributing 0 dB to the overall gain of the system and the integrator produces -20dB per decade thus:

$$| G(s)_{\text{OL}}|_{\text{dB}} = 20log(\frac{K_\text{P} \cdot G_{\text{INV}}}{L_\text{i}}) + 0 - 20log(\omega) \qquad (18)$$

At the cross over frequency ω_c the gain of the open loop system is $| G(s)_{\text{OL}}|_{\text{dB}} = 0$, therefore:

$$0 = 20log(\frac{K_\text{P} \cdot G_{\text{INV}}}{L_\text{i}}) - 20log(\omega_c) \qquad (19)$$

$$20log(\frac{K_\text{P} \cdot G_{\text{INV}}}{L_\text{i}}) = 20log(\omega_c) \qquad (20)$$

Thus:

$$K_\text{P} = \omega_c \cdot \frac{L_\text{i}}{G_{\text{INV}}} = \frac{2 \cdot \pi \cdot F_\text{s}}{3} \cdot \frac{L_\text{i}}{G_{\text{INV}}} \qquad (21)$$

3.1   Proportional Gain for PWM - 2 Level (2L) and 3L NPC

The static gain for the case of the two level inverter using PWM is $G_{\text{INV}} = \frac{V_{\text{dc}}}{2}$, therefore the proportional gain is finally:

$$K_{\text{PWM}} = \frac{4 \cdot \pi \cdot F_\text{s}}{3} \cdot \frac{L_\text{i}}{V_{\text{dc}}} \qquad (22)$$

According to Yazdani [5] the average model of the 3 Level NPC topology is the same as the 2 level PWM inverter topology, model that was illustrated in Eqs. (4) and (5). It follows that the gain calculated by the expression (22) is valid for both topologies, 2L PWM and 3L NPC PWM.

3.2   Proportional Gain for Space Vector Modulation (SVM) – 2L

Holmes in [11] and Mohan in [10] stated that the maximum amplitude of the output phase voltage (for a waveform measured in reference to the center point of the DC bus) for a sinusoidal reference voltage at one frequency is given by:

$$(v_{\text{alz}})_{\text{PWM}} = \frac{v_{\text{DC}}}{2} \qquad (23)$$

In case of SVM, the maximum amplitude of the output phase voltage is limited within the inner circle of the hexagon, in order to do not exceed the carried period, and it is given by [10]:

$$(v_{\mathrm{alz}})_{\mathrm{SVM}} = \frac{v_{\mathrm{DC}}}{\sqrt{3}} \tag{24}$$

Just by inspection and comparison we can conclude that the inverter gain in case of SVM is $G_{\mathrm{INV}} = \frac{v_{\mathrm{DC}}}{\sqrt{3}}$ and therefore the proportional gain in case of space vector modulation is given by:

$$K_{\mathrm{SVM}} = \frac{2\sqrt{(3)} \cdot \pi \cdot F_s}{3} \cdot \frac{L_i}{V_{\mathrm{dc}}} \tag{25}$$

3.3 Integral Gain

If it is wanted to add an integral term with and integrator in parallel with the proportional controller, the following expression is suggested for the integral gain:

$$K_{\mathrm{I}} = K_{\mathrm{P}} \cdot F_{\mathrm{S}} \cdot \frac{\pi}{3} \cdot \frac{1}{60} \tag{26}$$

The integral gain has been calculated in such a way that the cross over frequency and the phase margin are just determined by the proportional gain in Eq. (21). In other words, the proportional gain calculated by (26) does not introduce any phase lag at the desired cross over frequency.

4 Experimental Verification

In order to test the validity of the formulas (21) and (26) a Grid-Tie connected inverter was tested in a experimental setup similar to the scenario depicted in Fig. 1 and the parameters in Table 1. Figure 3 shows a picture of the hardware utilized during the experiment. The Grid-Tie connected converter is operated as STATCOM for injection of reactive power to the power system. The STATCOM is meant with a self-supporting DC-Bus, therefore it only absorbs from the power system the active power at the fundamental frequency necessary to compensate for the internal VSI losses. Invoking the instantaneous power theory [5], the active and reactive power that the power converter exchanges with the grid can be expressed by the following equations:

$$P(t) = \frac{3}{2} \cdot (v_{\mathrm{G}\alpha}(t) \cdot i_\alpha(t) + v_{\mathrm{G}\beta}(t) \cdot i_\beta(t)) \tag{27}$$

$$Q(t) = \frac{3}{2} \cdot (-1 \cdot v_{\mathrm{G}\alpha}(t) \cdot i_\beta(t) + v_{\mathrm{G}\beta}(t) \cdot i_\alpha(t)) \tag{28}$$

If we set P_{Ref} and Q_{Ref} as the active and reactive power that the Grid-Tie connected converter should inject into the grid, based on (27) and (28), the current references can be calculated as in (29) and (30).

Table 1. Experimental setup parameters

Description	Value
Power System	400 V (L-L), 50 Hz
Power Transformer	12.5 kVA, Imp. 3.3%, X/R = 2
Coupling Inductor	$L_i = 100\,\mu\text{H}$, $R_i = 1.6\,\text{m}\Omega$
Ripple Filter	$C_F = 300\,\mu\text{F}$, $R_F = 45\,\text{m}\Omega$
Modulation Method	PWM
Switching Frequency F_s	20 kHz
Current Controller Sampling Frequency	40 kHz
DC bus Voltage/Capacitor	750 V/8 mF

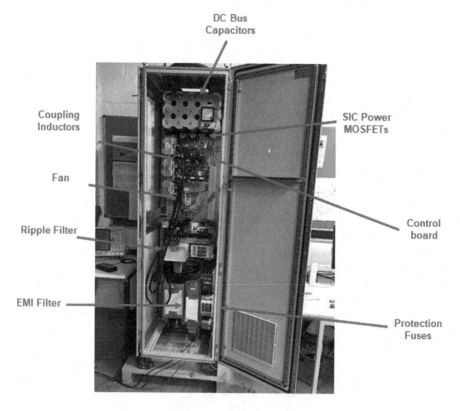

Fig. 3. Measured current of the grid-tie connected converter where the current controller was using the gains suggested in (21) and (26)

$$i_{\alpha ref}(t) = \frac{2}{3} \cdot \frac{v_{G\alpha}}{v_{G\alpha}^2 + v_{G\beta}^2} \cdot P_{\text{Ref}} + \frac{2}{3} \cdot \frac{v_{G\beta}}{v_{G\alpha}^2 + v_{G\beta}^2} \cdot Q_{\text{Ref}} \qquad (29)$$

$$i_{\beta ref}(t) = \frac{2}{3} \cdot \frac{v_{G\beta}}{v_{G\alpha}^2 + v_{G\beta}^2} \cdot P_{\text{Ref}} - \frac{2}{3} \cdot \frac{v_{G\alpha}}{v_{G\alpha}^2 + v_{G\beta}^2} \cdot Q_{\text{Ref}} \qquad (30)$$

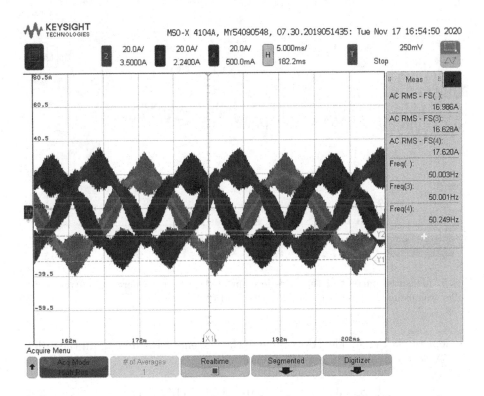

Fig. 4. Measured current of the grid-tie connected converter where the current controller was using the gains suggested in (21) and (26)

The complete system is controlled by a Texas Instrument microcontroller TMS320F28379DZWTT. The current controller is implemented in the microcontroller using the gains proposed in Sect. 3 with the parameters on Table 1. The digital implementation is carried out by transforming the continuous time domain current controller into a discrete time domain current controller using the bilinear transformation. The reference for the DC Bus voltage is set to 750 V, and the Q_{Ref} is set to 12 kVar. The measured currents during the experiment

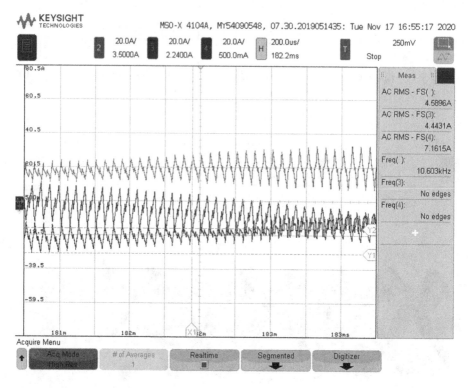

Fig. 5. Measured current of the grid-tie connected converter where the current controller was using the gains suggested in (21) and (26)

are depicted in Fig. 4. The RMS value of the currents measured are in the order of 17 A, leading to a total apparent power of:

$$S = 3 \cdot V \cdot I = 3 \cdot 230V \cdot 17A = 11.7kVA. \tag{31}$$

Due to the fact that the Grid-Tie connected converter uses SiC power MOS-FETs, the VSI internal losses are very small and the total apparent power injected by the converter is very close to the reference of 12 kVar set for Q_{Ref}. The current is zoomed in Fig. 5 and Fig. 6 in order to appreciated the current ripple of the injected currents. The current ripple is in the range of 20 A peak to peak.

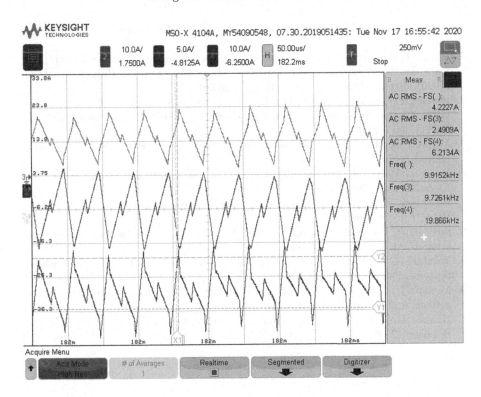

Fig. 6. Measured current of the grid-tie connected converter where the current controller was using the gains suggested in (21) and (26)

5 Conclusion

This paper has shown the derivation of an analytical formula to calculate the proportional gain of a current controller for a three PWM converter. The analytical formula is easy to follow; it only depends on the parameters of the three phase PWM converter. The idea behind the formula is to guarantee always a gain that maintains the inner current loop stable with a controller that provides a fast response. Furthermore, the formula can be used on an adaptive current controller when a nonlinear coupling inductivity is used. When the last is the case, based on the current reference at certain moment and a loop-up table with the L-I characteristic of the coupling inductivity, the proportional gain can be recalculated at all times for new current references feed to the current controller.

References

1. Kazmierkowski, M.P., Malesani, L.: Current control techniques for three-phase voltage-source PWM converters: a survey. IEEE Trans. Ind. Electron. 691–703 (1998). https://doi.org/10.1109/TIA.2004.827452

2. Trzynadlowski, A.M.: Introduction to Modern Power Electronics. Wiley, Hoboken (2015)
3. Jahns, T.M., Sarlioglu, B.: The incredible shrinking motor drive: accelerating the transition to integrated motor drives. IEEE Power Electron. Mag. 7(3), 18–27 (2020). https://doi.org/10.1109/MPEL.2020.3011275
4. Holmes, D.G., Lipo, T.A., McGrath, B.P., Kong, W.Y.: Optimized design of stationary frame three phase AC current regulators. IEEE Trans. Power Electron. 24(11), 2417–2426 (2009)
5. Yazdani, A., Iravani, R.: Voltage-Sourced Converters in Power Systems. Wiley, Oxford (2010)
6. Buso, S., Mattavelli, P.: Digital control in power electronics. Synthesis Lect. Power Electron. 1, 1–229 (2015)
7. Blaabjerg, F.: Control of Power Electronic Converters and Systems. Academic Press, Amsterdam (2017)
8. Silva, G.J., Bhattachaiyya, S.P., Datta, A.: PID Controllers for Time-Delay Systems. Birkhäuser, Boston (2005)
9. Franklin, G.F., Powell, J.D., Emami-Naeini, A.: Feedback Control of Dynamic Systems. Pearson, New York (2019)
10. Mohan, N., Undeland, T.M., Robbins, W.P.: Power Electronics, Converters, Applications, and Design. Wiley, Hoboken (2003)
11. Holmes, D.G., Lipo, T.A.: Pulse Width Modulation for Power Converters. Wiley, Hoboken (2003)

Limits of Harmonic Power Recovery by Power Quality Conditioners in Three-Phase Three-Wire Systems Under Non-sinusoidal Conditions

Rodrigo Guzman Iturra and Peter Thiemann(✉)

South Westphalia University of Applied Sciences,
Lübecker Ring 2, 59494 Soest, Germany
{guzmaniturra.rodrigozenon,thiemann.peter}@fh-swf.de
https://www4.fh-swf.de/de/home/

Abstract. Power quality conditioners were originally meant to provide protection to electrical loads connected to the power system against power quality disturbances. Such protection feature brings mostly intangible benefits that are difficult to quantify in monetary terms. This is the reason why some providers of power quality solutions focus more on promote energy savings benefits rather than emphasize the protection feature at the time of advertising their products. Sometimes, the energy savings are exaggerated and inflated with the aim to present a more convincing argument to the customers about why they should acquire a particular solution. This technical paper present presents two formulas that determine the theoretical maximum energy savings that can be achieved when a power quality conditioner targets current harmonics within an industrial facility. In particular, the formulas predict the maximum amount of harmonic active power that can be recovered by power quality conditioners (e.g. harmonic active power filter) in a three-wire three-phase system that contains linear and nonlinear loads. The upper bound of the harmonic active power is the total harmonic apparent power. The upper bound is given in function of the Total Harmonic Distortion of the current and the voltage measured at the point of common coupling and total apparent power of the loads.

Keywords: Active power filter · Energy savings · Harmonic active · Power power quality conditioner

1 Introduction

Due to the huge proliferation of loads of the nonlinear type, nowadays the most common voltage distortions encounter in power systems are related with harmonics, inter-harmonics, notching and noise [1,2]. Power quality solutions as passive harmonic filters and active power filters were originally meant to protect

© ICST Institute for Computer Sciences, Social Informatics and Telecommunications Engineering 2021
Published by Springer Nature Switzerland AG 2021. All Rights Reserved
J. L. Afonso et al. (Eds.): SESC 2020, LNICST 375, pp. 15–29, 2021.
https://doi.org/10.1007/978-3-030-73585-2_2

electrical devices from waveform distortions in the supplied voltage. Economically speaking, due to their protective feature, power quality solutions produce mostly intangible benefits to a particular industrial facility. Those savings are related with the minimization of production down-times and the damage reduction with the consequent lifetime increase of the electrical devices [2]. Indeed, it is very difficult to assign a hard value to such savings and thus they are considered as soft savings. This is the reason why, presently many of the power filter suppliers or representatives choose as sales strategy to promote the energy cost reduction rather than the protection feature [3]. Due to the economical challenges imposed over the current times, it is becoming every single time more and more difficult to get financial resources allocated to maintenance and improvement of existing processes. Therefore, power quality solutions that advertise the return of the investment in some finite amount of time makes a particular power quality conditioner device easier to sell. The former sales strategy offers an easy justification that pursuits to convince controllers or financial officers in charge of the purchase approval to accept the acquisition of a particular power quality conditioner. Consequently, many vendors of harmonic mitigation solutions embellish or inflate the energy savings that can be achieved. Sometimes energy savings on the range of 20% to 30% are claimed, when the true is that the energy savings are much smaller [4].

Many power filters vendors use the kVA instead of kW as method to make claims about significant energy reductions [3]. Equipment that improves or reduce the THD_U or THD_I can reduce the kVA demanded but have very little effect on the kWs consumed by the facility from the utility point of view. Real energy saving means a reduction on the real kW or kWh that a particular industrial facility demands. The energy required to do actual work (e.g. mechanical, heating, etc.) cannot be eliminated from the electrical costs unless a second energy source is installed locally [4]. Thus, if a device states to save energy on a power system, the only energy that can be saved is the energy wasted in losses through the power system. An important part of such losses are the power line losses. The former arise out due to the power dissipation produced in the line resistance due to the flow of fundamental and harmonics currents from the utility to the customer facility through the feeder and the distribution transformer. In this paper the focus will be solely on the power line losses due to harmonics currents. The power dissipated in the line resistance due to the flow of current harmonics is largely due to the harmonic active power (P_H) produced by the nonlinear loads/power converters [12]. The humble goal of this paper is to provide an easy to apply formula to estimate the amount of energy that can be saved if a particular power quality conditioner (e.g. harmonic power filter) processes or transforms the harmonic active power produced by nonlinear loads. In particular, in this paper the formula that determines the upper bound of the harmonic power that can be processed is given in function of the THD_U, THD_I and the total apparent power S_e of the loads measured at the point of common coupling (PCC) of the facility. Moreover, based on some practical experience, a very likely range where the true value of the harmonic power resides is given in function the upper bound.

2   Studied Power System

2.1   Scenario Definition

In order to study the energy savings potential due to the processing of current harmonics, some power definitions and the establishment the electrical quantities that are measured on a particular power system become necessary. Figure 1 shows the sketch of the studied circuit where three-phase symmetrical voltages with harmonic background distortion supply the power system through a distribution transformer [5]. The voltage at the secondary of the transformer is $U_1 = 230\,\text{V}$ RMS, 50 Hz with a background distortion at the fifth harmonic U_5 250 Hz and seven harmonic U_7 350 Hz. The impedance of the distribution feeder and transformer referred to the secondary of the transformer is represented by the equivalent components R_S and L_S [2]. The loads are balanced and consist on linear loads with ohmic inductive-behavior represented by the passive electrical elements R_L and L_L (e.g. induction motors) and nonlinear loads (NLV) (e.g. three phase passive diode rectifiers with smoothing DC-Link capacitor). The equivalent circuit can be seen in Fig. 2.

2.2   Voltages and Currents at the Coupling Point

In presence of nonlinear loads in the power system, the voltage at the PCC (U_N) contains a component at the fundamental u_{x1} with fundamental frequency $w_1 = 2 \cdot \pi \cdot f_1$ and harmonic voltages u_{xh} with frequencies at integers multiples of the fundamental frequency (possible also non-integers when interharmonics are considered) leading to harmonic frequencies $w_h = 2 \cdot \pi \cdot h \cdot f_1$. Therefore, the voltage at the PCC referred to the neutral point N can be mathematically written as in (1) to (3) [6]:

$$u_{nu} = u_{u1} + u_{uh} = \sqrt{2} \cdot U_1 \cdot \sin(\omega_1 t + \alpha_1) + \sum_{h=2}^{\infty} \sqrt{2} \cdot U_h \cdot \sin(h\omega_1 t + \alpha_h) \quad (1)$$

$$u_{nv} = u_{v1} + u_{vh} = \sqrt{2} \cdot U_1 \cdot \sin\left(\omega_1 t + \alpha_1 - \frac{2\pi}{3}\right)$$
$$+ \sum_{h=2}^{\infty} \sqrt{2} \cdot U_h \cdot \sin\left(h\omega_1 t + \alpha_h - h\frac{2\pi}{3}\right) \quad (2)$$

$$u_{nw} = u_{w1} + u_{wh} = \sqrt{2} \cdot U_1 \cdot \sin\left(\omega_1 t + \alpha_1 + \frac{2\pi}{3}\right)$$
$$+ \sum_{h=2}^{\infty} \sqrt{2} \cdot U_h \cdot \sin\left(h\omega_1 t + \alpha_h + h\frac{2\pi}{3}\right) \quad (3)$$

Where the terms expressed with capital letters U_h denote the RMS voltage quantity of the individual component. The total RMS voltage U_e of each phase

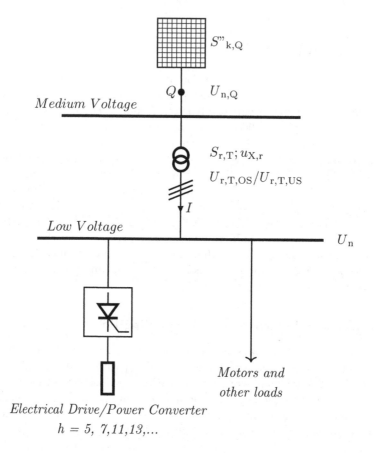

Fig. 1. Industrial power grid with linear and nonlinear loads connected to a utility power system through a power transformer.

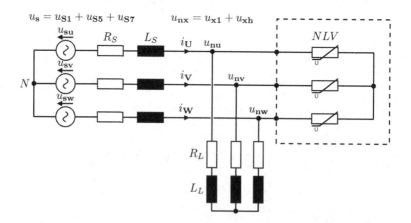

Fig. 2. Equivalent circuit of the power system depicted in Fig. 1.

considering all the components (fundamental and harmonics) was defined by Emanuel in [7] following the approach of Buchholz [8] and can be calculated as:

$$U_e = \sqrt{U_1^2 + U_H^2} = \sqrt{U_1^2 + \sum_{h=2}^{\infty} U_h^2} \tag{4}$$

Where the harmonic voltage contribution is:

$$U_H = \sqrt{\sum_{h=2}^{\infty} U_h^2} \tag{5}$$

The currents draw from the power system will have the same frequency components as the voltages at the PCC, thus the three phase currents can be expressed as in (6) to (8):

$$i_u = i_{u1} + i_{uh} = \sqrt{2} \cdot I_1 \cdot \sin(\omega_1 t + \beta_1) + \sum_{h=2}^{\infty} \sqrt{2} \cdot I_h \cdot \sin(h\omega_1 t + \beta_h) \tag{6}$$

$$i_v = i_{v1} + i_{vh} = \sqrt{2} \cdot I_1 \cdot \sin(\omega_1 t + \beta_1 - \frac{2\pi}{3})$$
$$+ \sum_{h=2}^{\infty} \sqrt{2} \cdot I_h \cdot \sin(h\omega_1 t + \beta_h - h\frac{2\pi}{3}) \tag{7}$$

$$i_w = i_{w1} + u_{wh} = \sqrt{2} \cdot I_1 \cdot \sin(\omega_1 t + \beta_1 + \frac{2\pi}{3})$$
$$+ \sum_{h=2}^{\infty} \sqrt{2} \cdot I_h \cdot \sin(h\omega_1 t + \beta_h + h\frac{2\pi}{3}) \tag{8}$$

Where capital letters I_h denote the current RMS quantity of the individual frequencies. The effective current of each phase considering the fundamental and harmonic frequencies can be calculated with an expression similar to (4) leading to:

$$I_e = \sqrt{I_1^2 + I_H^2} = \sqrt{I_1^2 + \sum_{h=2}^{\infty} I_h^2} \tag{9}$$

$$I_H = \sqrt{\sum_{h=2}^{\infty} I_h^2} \tag{10}$$

The total harmonic voltage and current distortion are defined as [9]:

$$THD_U = \frac{U_H}{U_1} = \sqrt{(\frac{U_e}{U_1})^2 - 1} \tag{11}$$

$$THD_I = \frac{I_H}{I_1} = \sqrt{(\frac{I_e}{I_1})^2 - 1} \tag{12}$$

3 Apparent Power Resolution for Nonsinusoidal Three-Wire Three Phase System

Considering that all the loads are connected to the PCC, any potential harmonic power that could processed and recovered later by and active power filters depends necessarily of the total voltage U_e that supplies the loads and the current I_e that such loads draw from the main power system. Therefore, the starting point for the calculation of the harmonic power limit that can be recovered is the total apparent power of the loads.

3.1 Total Apparent Power

Equations (4) and (9) lead to the total apparent power:

$$S_e = 3 \cdot U_e \cdot I_e \tag{13}$$

The apparent power can be squared and expanded as in [10]:

$$
\begin{aligned}
S_e^2 &= 9 \cdot U_e^2 \cdot I_e^2 = 9 \cdot (U_1^2 + U_H^2) \cdot (I_1^2 + I_H^2) \\
&= (U_1^2 \cdot I_1^2) + (U1^2 \cdot I_H^2) + (U_H^2 \cdot I_1^2) + (U_H^2 \cdot I_H^2) \\
&= (9 \cdot U_1^2 \cdot I_1^2) + (9 \cdot U_1^2 \cdot I_H^2) + (9 \cdot U_H^2 \cdot I_1^2) + (9 \cdot U_H^2 \cdot I_H^2) \\
&= (9 \cdot S_1^2) + (9 \cdot D_I^2) + (9 \cdot D_U^2) + (9 \cdot S_H^2) \\
&= (3 \cdot S_1)^2 + (3 \cdot D_I)^2 + (3 \cdot D_U)^2 + (3 \cdot S_H)^2 \\
&= (S_{e1})^2 + (D_{eI})^2 + (D_{eU})^2 + (S_{eH})^2
\end{aligned}
\tag{14}
$$

Equation (14) can be written as:

$$S_e^2 = S_{e1}^2 + S_{eN}^2 \qquad [V^2 \cdot A^2] \tag{15}$$

The first term in (15) is the square of the fundamental (50 Hz) apparent power S_{e1}:

$$S_{e1} = 3 \cdot S_1 = 3 \cdot U_1 \cdot I_1 \tag{16}$$

$$S_{e1} = 3 \cdot \sqrt{(P_1)^2 + (Q_1)^2} \tag{17}$$

$$S_{e1} = 3 \cdot \sqrt{\{U_1 \cdot I_1 \cdot cos(\beta_h - \alpha_h)\}^2 + \{U_1 \cdot I_1 \cdot sin(\beta_h - \alpha_h)\}^2} \qquad [V \cdot A] \tag{18}$$

The second term in (15) is the square of the nonfundamental apparent power S_eN, defined as following [11]:

$$S_{eN} = \sqrt{(D_{eI})^2 + (D_{eU})^2 + (S_{eH})^2} \qquad [V \cdot A] \tag{19}$$

Where:

$$D_{eI} = 3 \cdot D_I = 3 \cdot U_1 \cdot I_H \qquad [V \cdot A] \tag{20}$$

is the total current distortion power. The distortion power results from the product of the voltage at the fundamental and currents at different harmonic frequencies. The expression contains only oscillatory terms with an average value of zero, therefore D_{eI} produces just reactive power and its units is fundamentally Var. Most of the times, the distortion power D_{eI} makes the largest contribution to the magnitude of S_{eN}. Moreover:

$$D_{eU} = 3 \cdot D_U = 3 \cdot U_H \cdot I_1 \qquad [V \cdot A] \qquad (21)$$

is the total voltage distortion power. The voltage distortion power is the product of harmonic voltages at different frequencies and the current at the fundamental frequencys. It follows that D_{eU}, as D_{eI}, produces only reactive power leading to Var as unit. Finally, the smallest of all the terms in (3.5) is the total harmonic apparent power S_{eH}:

$$S_{eH} = 3 \cdot S_H = 3 \cdot U_H \cdot I_H \qquad [V \cdot A] \qquad (22)$$

The harmonic apparent power can be characterized by two important components [11]:

$$S_{eH}^2 = P_H^2 + D_{eH}^2 \qquad [V^2 \cdot A^2] \qquad (23)$$

We recall that the active power P_x is equal to the average value of the instantaneous power $p_x = v_x \cdot i_x$ over a time window for the averaging equal to one grid voltage period. It is granted that if we consider a three phase system, the contribution of each phase to the active power needs to be considered. With the last arguments on mind, the total harmonic active power P_H is formally defined as [12]:

$$P_H = \int_0^{\frac{1}{f_1}} \left\{ p_{uh} + p_{vh} + p_{wh} \right\} dt \qquad (24)$$

$$P_H = \int_0^{\frac{1}{f_1}} \left\{ (u_{uh} \cdot i_{uh}) + (u_{vh} \cdot i_{vh}) + (u_{wh} \cdot i_{wh}) \right\} dt \qquad (25)$$

$$P_H = 3 \cdot \sum_{h=2}^{\infty} U_H \cdot I_H \cdot cos(\beta_h - \alpha_h) \qquad [W] \qquad (26)$$

Which has an average value different from zero due to the product of currents and voltages at the same frequency. Actually, this one is the only one component of the apparent power than can be processed later to recover energy from the harmonics present in the power system. The term D_{eH} is defined as harmonic distortion power and results from the product of harmonic voltages and currents at different frequencies. The consequence is that D_{eH} contains only oscillatory terms and thus produces only reactive power:

$$D_{eH} = \sqrt{(S_{eH})^2 - (P_H)^2} \qquad [V \cdot A] \qquad (27)$$

3.2   Apparent Power Components as Functions of THD

The different components that form the total apparent power (14) can be expressed as functions of the total harmonic voltage and current distortion THD_U and THD_I [12]. This can be achieved by multiplying and dividing each term by the factor S_{e1}. For instance, take the total current distortion power D_{eI} in equation (20):

$$D_{eI} = 3 \cdot U_1 \cdot I_H = S_{e1} \cdot \frac{3 \cdot U_1 \cdot I_H}{3 \cdot U_1 \cdot I_1} = S_{e1} \cdot \frac{I_H}{I_1} = S_{e1} \cdot THD_I \qquad (28)$$

A similar procedure can be carried out with the total voltage distortion power D_{eU} and the harmonic apparent power S_{eH}:

$$D_{eU} = 3 \cdot U_H \cdot I_1 = S_{e1} \cdot \frac{3 \cdot U_H \cdot I_1}{3 \cdot U_1 \cdot I_1} = S_{e1} \cdot \frac{U_H}{U_1} = S_{e1} \cdot THD_U \qquad (29)$$

$$S_{eH} = 3 \cdot U_H \cdot I_H = S_{e1} \cdot \frac{3 \cdot U_H \cdot I_H}{3 \cdot U_1 \cdot I_1} = S_{e1} \cdot \frac{U_H}{U_1} \cdot \frac{I_H}{I_1} = S_{e1} \cdot THD_U \cdot THD_I \quad (30)$$

Replacing (28), (29) and (30) in (19) leads to:

$$S_{eN} = S_{e1} \cdot \sqrt{(THD_I)^2 + (THD_U)^2 + (THD_U \cdot THD_I)^2} \qquad [V \cdot A] \qquad (31)$$

Substituting (31) in (15) results in:

$$S_e = S_{e1} \cdot \sqrt{1 + (THD_I)^2 + (THD_U)^2 + (THD_U \cdot THD_I)^2} \qquad [V \cdot A] \quad (32)$$

4   Determination of the Limits or Bounds for P_H

From the discussion at chapter 3, it was concluded that the only single power component from the nonfundamental apparent power that do not produce reactive power is the harmonic active power P_H. Therefore, the limit of the power that can be recovered from harmonics is going to be determined by the magnitude of P_H. From (26), it can be seen that this quantity depends on the magnitude of each harmonic voltage, current harmonic and the phase angle difference between them. To the best of our knowledge, it is not possible to express the power P_H returned by the nonlinear loads to the power system in terms of THD_U and THD_I at the point of common coupling directly. However, Eq. (23) relates P_H with S_{eH}, here rewritten in a different way:

$$S_{eH} = \sqrt{P_H^2 + D_{eH}^2} \qquad [V \cdot A] \qquad (33)$$

Equation (30) gives a defined value for S_{eH} in terms of THD_U and THD_I. The question is how to related the limit or upper bound of P_H knowing the value of S_{eH}, issue that is not trivial. Let us to define the following terms:

$$a_1 = P_H^2 \tag{34}$$

$$a_2 = D_{eH}^2 \tag{35}$$

Thus (33) becomes:

$$S_{eH} = \sqrt{a_1 + a_2} \tag{36}$$

It is straight forward to demonstrate the following inequality [13]:

$$\sqrt{a_1 + a_2} \leq \sqrt{a_1} + \sqrt{a_2} \tag{37}$$

Replacing (34), (35) and (36) in the inequality (37), the following expression is obtained:

$$S_{eH} \leq \sqrt{P_H^2} + \sqrt{D_{eH}^2} \tag{38}$$

Moreover, the following mathematical expression holds [14][15]:

$$\sqrt{a^2} = |a| \tag{39}$$

Using (39), Eq. (38) becomes:

$$S_{eH} \leq |P_H| + |D_{eH}| \tag{40}$$

It follows:

$$|P_H| \geq S_{eH} - |D_{eH}| \tag{41}$$

The value of S_{eH} is by the definition given in Eq. (22) always a positive number [11]. Therefore the maximum absolute value that P_H can reach will necessary be equal to S_{eH} when D_{eH} is equal to zero. The former will be the most optimistic scenario of the amount of harmonic power that can be transformed. Indeed having a network with $D_{eH} = 0 \, Var$ is not a practical case, due to the fact that in general $S_{eH} > D_{eH} > P_H$, so we are in the safe side if we claim that the upper bound for P_H is the S_{eH} magnitude. Therefore, any potential energy saving that can be achieved by transforming harmonic active power is going to be upper bounded by S_{eH}. If we aim to calculate the energy savings potential in percentage that can be achieved by processing or transforming harmonic active power, the ratio $\frac{S_{eH}}{S_{e1}}$ can be used. Equation (30) leads to the following $\frac{S_{eH}}{S_{e1}}$ ratio expression:

$$\frac{S_{eH}}{S_{e1}} = THD_U \cdot THD_I \tag{42}$$

The former equation implies that the harmonic active power recovery limit related to the fundamental apparent power $S_{e1} = 3 \cdot S_1$ is going to be bounded by the product of $THD_U \cdot THD_I$. For instance if in a particular industrial facility the THD_U is 8% (which is already the maximum limit impose by the standard IEC 61000-2-4 for facilities class 2) and the THD_I is 60% (which is a typical THD_I in many cases [12]), the energy savings potential in percentage related to the fundamental apparent power of the facility is bounded by:

$$\frac{S_{eH}}{S_1} = \frac{8}{100} \cdot \frac{60}{100} \cdot 100\% = 4.8\% \tag{43}$$

Most of the times the fundamental apparent power $S_{e1} = 3 \cdot S_1$ cannot be measured without a power analyzer that separates the different frequencies contained in the measured voltages and currents through Fourier Analysis. Most of the times what is available is the TRUE RMS value of currents and voltages that can be acquired easily with typical multimeters. The TRUE RMS value considers several frequency components in a range 45 Hz to ten's of kHz all at once. Instruments measuring the TRUE RMS are for instance the FLUKE 175 [16] or the power meter SIEMENS SENTRONPAC 3200 [17]. If the former is the case, the electrical quantity that is available is the total apparent power S_e. In order to related S_{eH} to the total apparent quantity S_e, let us square Eq. (30) to obtain a new equation:

$$S_{eH}^2 = S_{e1}^2 \cdot (THD_U \cdot THD_I)^2 \tag{44}$$

If (44) is divided by the square of (32), it follows automatically that:

$$\frac{S_{eH}^2}{S_e^2} = \frac{(THD_U \cdot THD_I)^2}{1 + (THD_I)^2 + (THD_U)^2 + (THD_U \cdot THD_I)^2} \tag{45}$$

$$\frac{S_{eH}}{S_e} = \sqrt{\frac{(THD_U \cdot THD_I)^2}{1 + (THD_I)^2 + (THD_U)^2 + (THD_U \cdot THD_I)^2}} \tag{46}$$

If the energy savings potential due to harmonic active power is calculated using (46) with $THD_U = 8\%$ and $THD_I = 60\ \%$, as in Eq. (43), the limit of the harmonic active power that can be recovered related to the total apparent power of the facility is given by:

$$\frac{S_{eH}}{S_e} = \sqrt{\frac{(0.08 \cdot 0.6)^2}{1 + (0.6)^2 + (0.08)^2 + (0.08 \cdot 0.6)^2}} = 4.1\% \tag{47}$$

In the above examples the standard IEC 61000-2-4 was referenced to set the THD_U to 8%, where a total voltage distortion of 8% is the maximum value allowed for an industrial facility. Nonetheless, the standard IEC 61000-2-4 do not define a maximum value for the THD_I. The value of THD_I varies typically between 5% and 120% [12]. Furthermore, if the industrial facility needs to comply with the standard IEEE 519-1992 which impose more severe restrictions on the THD_U and THD_I than the IEC 61000-2-4 norm, the harmonic power recovery potential is reduced. The IEEE Standard 519-1992 defines a maximum THD_U of 5% for low voltage power systems (<69 kV). The IEEE Standard 519-1992 defines the maximum TDD_I (TDD_I is equal to THD_I when the facility operates almost at 100% load) in function of the ratio of the maximum short circuit current at the PCC I_{SC} and the maximum current load at the fundamental frequency I_L [2]. If the maximum allowed $TDD_I= 20\%$ is taken, the limit of the harmonic active power that can be recovered according to (46) is reduced to:

$$\frac{S_{eH}}{S_e} = \sqrt{\frac{(0.05 \cdot 0.2)^2}{1 + (0.2)^2 + (0.05)^2 + (0.05 \cdot 0.2)^2}} = 0.97\% \qquad (48)$$

5 Simulation Validation

In order to confirm the theoretical derivations stated in Sects. 3 and 4, simulations are carried out. The simulation is performed in Simulink using elements of the Simscape Electrical Specialized Library. The scenario simulated is similar to the one depicted in Fig. 1 with the parameters stated in Table 1. The behavior of the voltages at PCC and the currents drawn from the power system can be seen in Fig. 3 and Fig. 4 respectively.

Table 1. Simulation parameters

Description	Value
Power System Fundamental	$U_1 = 230\,V,\ 50\,Hz$
Voltage Background Distortion	$U_5 = 4.6\,V,\ 250\,Hz$
	$U_7 = 2.07\,V,\ 350\,Hz$
Power Transformer	2 MVA, Imp. 5%, X/R = 20
Linear Load (Resistive-Inductive)	$P_1 = 700\,kW;\ Q_1 = 100\,kVAR$
Nonlinear Load	700 kW, Passive Diode Bridge + Smoothing Capacitor

The components up to the 13th harmonic of the measured voltages u_{xn} at the coupling point and the currents i_x drawn from the power systems, with x $x \in u,v,w$ can be seen in Tables 2 and 3 respectively. The effective value of the voltage and current U_E and I_E are also written in the tables. Based on the results of Tables 2 and 3 and Eqs. (13) to (32), it is possible to calculate each of the components of the apparent power under nonsinusoidal conditions leading to the results in Table 4.

The real absolute harmonic active power that can be recovered or transformed is $P_H = 3712W$ measured in the simulation. Let us test the boundaries determined by the Eq. (42). Taking the $THD_U = 7.92\%$ and $THD_I = 26.73\%$ from Tables 2 and 3, the equation predicts the upper bound of the amount of harmonic active power that can be recovered as:

$$\frac{S_{eH}}{S_{e1}} = \frac{7.92}{100} \cdot \frac{26.93}{100} \cdot 100\% = 2.13\% \qquad (49)$$

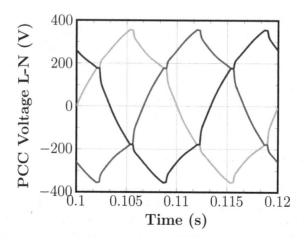

Fig. 3. Voltages at PCC u_{nu}, u_{nv}, u_{nw}.

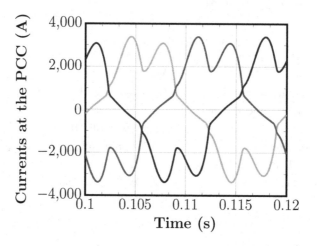

Fig. 4. Currents drawn from the power system i_u, i_v, i_w.

Table 2. Measured voltages [V] (Magnitudes in RMS)

h	u	v	w
1	$227.6 \angle 177.7°$	$227.6 \angle 57.7°$	$227.6 \angle -62.3°$
5	$14.82 \angle 191.8°$	$14.82 \angle -48.2°$	$14.82 \angle 71.8°$
7	$7.48 \angle -73.5°$	$7.48 \angle 166.5°$	$7.48 \angle 46.5°$
11	$3.8 \angle -33.8°$	$3.8 \angle 86.2°$	$3.8 \angle 206.2°$
13	$2.84 \angle 41.7°$	$2.84 \angle -78.3°$	$2.84 \angle 161.7°$
U_e	228.3	228.3	228.32
THD_U	7.92%	7.92%	7.92%

Table 3. Measured currents [A] (Magnitudes in RMS)

h	u	v	w
1	2002 ∠ 167.3°	2002 ∠ 47.3°	2002 ∠ -72.7°
5	481.54 ∠ 268.1°	481.54 ∠ −28.1°	481.54 ∠ 148.1°
7	211.34 ∠ -33°	211.34 ∠ −87°	211.34 ∠ 153°
11	72.11 ∠ -56.5°	72.11 ∠ 176.5°	72.11 ∠ −63.5°
13	45.6 ∠ 131.9°	45.6 ∠ −11.9°	45.6 ∠ -63.5°
I_e	2072	2072	2072
THD_I	26.73%	26.73%	26.73%

Table 4. Components of the apparent power calculated based on Tables 2 and 3

Quantity	Unit	Value
S_e	VA	1419100
S_1	VA	1367000
S_{eN}	VA	381160
$P_{e1} = 3 \cdot P_1$	W	1344500
$Q_{e1} = 3 \cdot Q_1$	Var	246760
D_{eI}	Var	365390
D_{eV}	Var	108260
S_H	VA	28939
P_H	W	3712
D_{eH}	Var	28700

If S_{eH} and S_{e1} are read directly from the simulation results in Table 4, the former ratio can be directly calculated as:

$$\frac{S_{eH}}{S_{e1}} = \frac{28939VA}{1367000VA} \cdot 100\% = 2.12\% \tag{50}$$

It is possible to observe the very good agreement between the results of (49) and (50). Moreover, Eq. (49) suggest that the upper bound (O) of harmonic active power that can be recovered ($D_{eH} = 0\ Var$) is:

$$]O(P_H) = S_{e1} \cdot cos(0) \cdot THD_U \cdot THD_I = 28939\,\text{W} \tag{51}$$

The bound predicted by Eq. (51) is clearly above the real $P_H = 3712W$ measured in simulation proving that Eq. (42) is correct. It is possible to infer that the real amount of harmonic active power that can be recovered will be in the range $0 < P_H < O(P_H)$. Probably a most realistic range for the amount of harmonic active power that can be recovered is between $0.1 \cdot O(P_H) < P_H < 0.5 \cdot O(P_H)$ due to the fact that in most cases D_{eH} is different from 0. On the other hand, Eq. (46) based on the total apparent power leads to the same results:

$$\frac{S_{\mathrm{eH}}}{S_{\mathrm{e}}} = \sqrt{\frac{(0.0792 \cdot 0.26)^2}{1 + (0.26)^2 + (0.0792)^2 + (0.0792 \cdot 0.26)^2}} = 2.03\% \qquad (52)$$

From the simulation results in Table 4, it is possible to calculate:

$$\frac{S_{\mathrm{eH}}}{S_{\mathrm{e}}} = \frac{28939VA}{1419100VA} \cdot 100\% = 2.04\% \qquad (53)$$

Equation (46) agrees very well with the simulation results as is demonstrated by the same output of equations (52) and (53). The amount of harmonic active power that can be recovered related to the total apparent power is:

$$O(P_{\mathrm{H}}) = S_{\mathrm{e}} \cdot cos(0) \cdot \sqrt{\frac{(THD_{\mathrm{U}} \cdot THD_{\mathrm{I}})^2}{1 + (THD_{\mathrm{I}})^2 + (THD_{\mathrm{U}})^2 + (THD_{\mathrm{U}} \cdot THD_{\mathrm{I}})^2}} \qquad (54)$$

$$O(P_{\mathrm{H}}) = 1419100 \, \mathrm{VA} \cdot cos(0) \cdot 2.03\% = 28935 \, \mathrm{W} \qquad (55)$$

Leading to the same result as Eq. (51) and proving the correctness of (46).

6   Conclusions

This paper has shown the derivation of two formulas (42) and (46) that determine the maximum amount of harmonic active power (P_{H}) that can be recovered in a three-wire three-phase system that contains linear and nonlinear loads. The harmonic active power recovery could be performed by a power conditioner device such as harmonic active power filter. The upper bound (O) of the harmonic active power $O(P_{\mathrm{H}})$ is S_{eH}. The upper bound is given in function of the Total Harmonic Distortion of the currents (THD_{I}) and the voltage (THD_{U}) measured at the point of common coupling (PCC), the fundamental apparent power S_{e1} and total apparent power S_{e} of the loads. Probably the realistic range of the real amount of P_{H} that can be recovered is $0.1 \cdot O(P_{\mathrm{H}}) < P_{\mathrm{H}} < 0.5 \cdot O(P_{\mathrm{H}})$. For example, if in a particular industrial facility, the THD_{U} is 8% (which is already at the maximum limit imposed by the standard IEC 61000-2-4 for facilities class 2) and the THD_{I} is 60% which is typical for many cases, the maximum saving potential in percentage by processing harmonic active power will be bounded by 4.10%. Where it is in highly probable that most realistic value will be between the 0.4% and 2.05%.

References

1. Abu-Rub, H., Malinowski, M., Al-Haddad, K.: Power Electronics for Renewable Energy Systems, Transportation and Industrial Applications, pp. 534–535. Wiley, Chichester (2014)
2. Singh, B., Chandra, A., Al-Haddad, K.: Power Quality: Problems and Mitigation Techniques, pp. 356–358. Wiley, Hoboken (2014)

3. Carnovale, D.J., Barchowsky, A., Groden, B.: Energy savings-realistic expectations for commercial facilities (2015). https://www.eaton.com/FR/FTC/buildings/KnowledgeCenter/WhitePaper2/index.htm. Accessed 15 Oct 2020

4. Carnovale, D.J., Hronek, T.: Power quality solutions and energy savings–what is real? Energy Eng. **106**, 26–50 (2009). https://doi.org/10.1080/01998590909509179. Fluke 175 True-RMS Digital Multimeter—Fluke. https://www.fluke.com/en-us/product/electrical-testing/digital-multimeters/fluke-175. Accessed 17 Sept 2020

5. Schlabbach J., Mombauer W.: Power Quality: Entstehung und Bewertung von Netzrückwirkungen, Netzanschluss erneuerbarer Energiequellen, Theorie, Normung und Anwendung von DIN EN 61000-3-2 (VDE 0838-2), DIN EN 61000-3-12 (VDE 0838-12), DIN EN 61000-3-3 (VDE 0838-3), DIN EN 61000-3-11 (VDE 0838-11), DIN EN 61000-2-2 (VDE 0839-2-2), DIN EN 61000-2-4 (VDE 0839-2-4), DIN EN 61000-4-7 (VDE 0847-4-7), DIN EN 61000-4-15 (VDE 0847-4-15), DIN EN 50160, DIN EN 61000-4-30 (VDE 0847-4-30), VDN-Technische Regeln zur Beurteilung von Netzrückwirkung: VDE, pp. 247–248 (2008)

6. Santoso, S.: Fundamentals of Electric Power Quality, 2009th edn., pp. 207–208. Published by Surya Santoso Through CreateSpace, Austin (2009)

7. Emanuel, A.E.: Summary of IEEE standard 1459: definitions for the measurement of electric power quantities under sinusoidal, nonsinusoidal, balanced, or unbalanced conditions. IEEE Trans. Ind. Appl. **40**(3), 869–876 (2004). https://doi.org/10.1109/TIA.2004.827452

8. Buchholz, F.: Die drehstrom-scheinleistung bei ungleichmassiger belastung der drei zweige. Licht und Kraft **2**, 9–11 (1922)

9. IEEE Standard Definitions for the Measurement of Electric Power Quantities Under Sinusoidal, Nonsinusoidal, Balanced, or Unbalanced Conditions, pp. 8–9 (2010)

10. Mohamadian, S., Shoulaie, A.: Comprehensive definitions for evaluating harmonic distortion and unbalanced conditions in three- and four-wire three-phase systems based on IEEE standard 1459. IEEE Trans. Power Deliv. **26**(3), 1774–1782 (2011). https://doi.org/10.1109/TPWRD.2011.2126609

11. Schulz, D.: Netzrückwirkungen-Theorie, Simulation, Messung und Bewertung. In: VDE-Schriftenreihe-Normen verständlich, vol. 115 (2004)

12. Emanuel, A.E.: Power Definitions and the Physical Mechanism of Power Flow, pp. 110–189. Wiley, Hoboken (2011)

13. Mathematics Stack Exchange, real analysis - square root of a sum? Bound? - Mathematics Stack Exchange. https://math.stackexchange.com/questions/318649/square-root-of-a-sum-bound/444640. Accessed 17 Sept 2020

14. Stewart, J.: Calculus: Concepts and Contexts. Cengage, Boston (2019)

15. Polianin, A.D., Manzhirov, A.V.: Handbook of Mathematics for Engineers and Scientists. Chapman & Hall/CRC, Boca Raton (2007)

16. Fluke 175 True-RMS Digital Multimeter—Fluke. https://www.fluke.com/en-us/product/electrical-testing/digital-multimeters/fluke-175. Accessed 17 Sept 2020

17. 7KM2112-0BA00-3AA0 - Industry Support Siemens. https://support.industry.siemens.com/cs/pd/107189?pdti=td&dl=en&lc=en-CN. Accessed 17 Sept 2020

Comparative Analysis and Validation of Different Modulation Strategies for an Isolated DC-DC Dual Active Bridge Converter

Sergio Coelho[✉], Tiago J. C. Sousa, Vitor Monteiro, Luis Machado, Joao L. Afonso, and Carlos Couto

Centro ALGORITMI, University of Minho, Campus de Azurém, Guimarães, Portugal
sergio.coelho@algoritmi.uminho.pt

Abstract. This paper presents a comparative analysis of different modulation techniques that can be applied to a dual active bridge (DAB) converter, validating and analyzing its performance with the realization of computational simulations.

A DAB converter is an isolated dc-dc topology with great applicability in the most diverse branches of power electronics, as is the case of energy storage systems, solid state transformers, power electronic traction transformers, and, more recently, dc or hybrid microgrids. In this sense, several strategies have been studied to mitigate circulating currents, expand the zero voltage switching operating range, and reduce reactive power, as well as semiconductor stress. One of the possible solutions to increase the efficiency of this dc-dc converter is to adopt specific modulation techniques, however, it is necessary to assess which one has a better cost-benefit ratio. Thus, this paper presents a comparative analysis between: (i) Duty-cycle modulation; (ii) Single phase shift (SPS); (iii) Dual phase shift (DPS); (iv) Extended phase shift (EPS); (v) Triple phase shift (TPS). Specifically, this comparative analysis aims to investigate the performance of a DAB converter when controlled by the aforementioned strategies and operating with a nominal power of 3.6 kW, a switching frequency of 100 kHz, and a transformation ratio of 2:1. Considering these operation parameters and by analyzing the obtained simulation results, it was shown that only SPS, DPS, and TPS modulation techniques are considered suitable for this particular case. Duty-cycle modulation presents time limitations during the power transfer, whilst EPS is more suitable for dynamic medium/high power applications since it is capable of transferring a certain power value in a short period of time.

Keywords: Dual active bridge converter · Duty-cycle modulation · Single phase shift · Dual phase shift · Extended phase shift · Triple phase shift

1 Introduction

The dual active bridge (DAB) converter was firstly introduced in [1] for the purpose of practical implementation in medium/high power applications, as is the case of aerospace

J. L. Afonso et al. (Eds.): SESC 2020, LNICST 375, pp. 30–49, 2021.
https://doi.org/10.1007/978-3-030-73585-2_3

systems. Since this converter is capable of providing galvanic isolation and operate in a high-frequency regime, its power density is necessarily higher and, consequently, its passive components are considered to have less volume and weight [2]. In this sense, according to the latest trends and technologic advances in the field of power electronics (digital signal processors, semiconductors based on GaN and SiC, etc.), the study and applicability of the DAB converter have been receiving special prominence in the interface with energy storage systems, in the design of on board battery chargers for electric vehicles (EV) [3], in the mitigation of power quality problems [4], in solid state transformers [5], in power electronic traction transformers [6] and, more recently, in medium/low voltage dc (or hybrid) power grids, such as smart microgrids [7, 8].

Given that the architecture of a DAB converter is considered quite simple and symmetric, in which two H-bridges are connected by a high-frequency transformer (Fig. 1), a great number of advantages are provided to this topology when compared to the remaining high-frequency dc-dc converters referred in the literature. The latter include the dual active half bridge and the full bridge forward topologies, covered in [9] and [10], respectively. Besides the bidirectional operation and the galvanic isolation, common to the three topologies, the soft-switching capacity is also considered through the entire power range, and there is the possibility of including more degrees of freedom in the control algorithms.

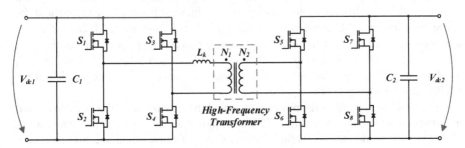

Fig. 1. Dual active bridge (DAB) converter.

However, to further increase the efficiency of this isolated dc-dc converter, new topologies derived from DAB have recently emerged, as is the case of the three-level DAB converter with five control degrees of freedom [11], the neutral point clamped DAB converter [12], the multilevel dc-dc DAB converter utilizing an *LCL* filter at the transformer side [13] and the dual bridge series resonant converter [14], which, compared to the traditional DAB, has as main features two resonant tanks and a tapped-transformer in order to increase the soft-switching range of the converter. In turn, in [15] the feasibility of connecting two DAB converters in parallel is studied, so that the efficiency of the system is optimized for different voltage and power demand levels.

Consequently, verifying the applicability of the DAB converter in low power systems, where the total losses must necessarily be as low as possible, new modulation techniques have been developed in order to regulate the current and voltage values of the system [16, 17]. Additionally, it is vital that a DAB converter assumes a dynamic behavior in the face of transient states and oscillations in the operating conditions. Therefore, in [18] a dynamic algorithm is implemented, capable of controlling each semiconductor independently, modifying their duty-cycle value in transient states. Such a measure would allow to eliminate current offsets (both the current that flows through the leakage inductance (i_{L_k}) and the magnetization current of the transformer) and, consequently, mitigate peak currents in the semiconductors. Similarly, in [19] the implementation of a linear quadratic regulator control based on linear matrix inequalities is considered to provide higher efficiency to the DAB converter when there is great uncertainty in the operating parameters of the system.

Increasing the efficiency of the DAB converter mainly involves the adoption of the latest generation semiconductors (SiC, GaN, etc.) and the reduction of the reactive power [20] and circulating current [21] values. However, in [22] issues related to possible transformer saturation and current peaks in systems that require a fast dynamic response are addressed, while in [23], is verified the attenuation of current stress in each of the semiconductors that make up the DAB converter architecture. In [24], a modulation strategy with 4 degrees of freedom is implemented with the goal of minimizing the total system losses. To this end, the necessary conditions to achieve soft-switching throughout the entire range of the DAB converter were considered and minimum-tank-current strategies were adopted in order to reduce the reactive power to zero value. To achieve the same objective, i.e., maximize the efficiency of the DAB converter, in [25] and in [26], a voltage offset is introduced to the dc blocking capacitors present on each side of the DAB, reducing, consequently, i_{L_k}.

Nonetheless, to regulate the power flow to and from each side of the DAB converter, the (traditional) single phase shift (SPS) modulation is often used. Such common use of this type of modulation is mainly due to its ease of implementation, however, high values of reactive power and circulating current are generated, thus making it difficult to implement zero voltage switching (ZVS) techniques. The expansion of the ZVS operating range is particularly relevant throughout the design and development of a DAB converter, which is why, in the literature, different methods and conditionings are studied to achieve this same objective. In [27] is presented a control technique capable of overcoming energy efficiency problems (among them the ZVS range limitations) that would arise in situations where the input voltage (v_{dc1}) does not match $n \bullet v_{dc2}$, with v_{dc2} being the DAB output voltage and n the transformation ratio of the high-frequency transformer. On the other hand, in [28] is verified the impact of dividing the leakage inductance (L_k) in the ZVS operating range. In turn, to expand this range, in [29], a control strategy for high-voltage applications is presented, in [30] a modulation for applications with wide input/output voltage is addressed and, lastly, in [31] the employability of a dual-transformer-based DAB converter is considered.

Among all the aforementioned methods, the ZVS operating range can also be expanded with the implementation of one of three phase shift modulation techniques: dual phase shift (DPS), extended phase shift (EPS) [32], and triple phase shift (TPS) [33]. These are based on the addition of more degrees of freedom to the system and seen as essential to mitigate energy efficiency problems, as is the case of semiconductor high-current stress [34]. Among the four main variants of the phase shift modulation (SPS, DPS, EPS, and TPS), the primary difference lies in the value assigned to each of the phase-lag angles, either between the two H-bridges (outer phase angle, D_0) or between the legs of each of the bridges (inner phase angles, D_1 and D_2, applied depending on the chosen modulation technique). The adoption of each one of these techniques, as mentioned above, will have a direct consequence on the efficiency presented by the DAB converter, however, the implementation difficulty will also be different. Normally, the value of D_0 controls the direction and the value of the transferred power to and from each side of the DAB converter, while D_1 and D_2 regulate and expand the ZVS range and minimize the circulating current, reactive power, and current peaks.

However, to mitigate faults, efficiency and reactive power problems, and possible power fluctuations, new modulation techniques, derived from the four above, have emerged. In this sense, in [35] and in [36] are presented derivations of the TPS and EPS modulation, both aiming to suppress the occurrence of transient dc bias. In [37] and [38], an improved/cooperative TPS modulation is presented, capable of mitigating dual-side circulating currents. A new DPS variant is considered in [39]. In [40] a unified phase shift modulation is studied, commonly used in dynamic systems, as is the case of electric traction drive systems. The latter is responsible for reducing peak currents during the moments that, e.g., fluctuations in the input voltage value occur. In [41], a flux control modulation is exposed, derived from SPS, which aims to keep the utilization of the transformer core constant on the whole load range during power transfer. Lastly, in [42] a novel hybrid current modulation is presented, based on triangular current modulation and trapezoidal current modulation techniques, which aims to reduce THD problems without increasing the stress caused in semiconductors and the RMS value of the DAB current.

Despite already existing comparative studies on the application of the aforementioned phase shift modulation techniques (SPS, DPS, EPS, and TPS) [43–46], no reference compares the four techniques simultaneously, representing the main contribution of this paper. In addition, it would be important to verify the behavior of the DAB converter when operating as a single active bridge (SAB), i.e., when only one H-bridge is switching, with the other adopting a passive behavior (current flowing through diodes). In this sense, this paper organizes as follows: Sect. 1 introduces the main ideas adjacent to this paper, presenting, at the same time, a review on the state-of-the-art. The operating principle of the system is referred in Sect. 2, while Sect. 3 describes each modulation technique applied to the DAB converter, explaining their advantages and disadvantages. Additionally, the obtained simulation results (performed with PSIM 9.1) are also compared in this section, observing current and voltage waveforms in the high-frequency transformer and in the resistive load, as well as the triggering gate signals of each semiconductor. Lastly, the main conclusions of this scientific article are presented.

2 Operating Principle of the System

As aforementioned, the adoption of the correct modulation strategy for the DAB converter will be translated into a great number of advantages in terms of energy efficiency. Thereby, in the first instance, it is crucial to define the operating characteristics of the system, so that, subsequently, each modulation technique can be analyzed and discussed in greater detail. The parameters of the DAB converter and respective values are shown in Table 1, however, it is important to note that they will remain constant throughout the simulations of each type of modulation. This measure was adopted to approximate the performed simulations to a real application case, where the physical components cannot be changed depending on the chosen control technique.

Among all the values presented in Table 1, it is important to highlight the values assigned to the switching frequency (f_{sw}) and to the L_k. The choice of such a high value for f_{sw} meets the recent trends in power electronics systems (SiC and GaN-based semiconductors) and enables the compaction of the system in case of possible implementation, allowing, consequently, the reduction in the volume and weight of inductors and capacitors. On the other hand, the value of L_k was selected in order to obtain the best compromise between the expansion of the ZVS range and the total losses of the DAB. As (1) demonstrates, the latter are the result of the sum of conduction losses (P_{cond}), switching losses (P_{sw}), losses in the transformer (P_{tr}) and, lastly, losses in L_k (P_{L_k}).

$$P_t = P_{cond} + P_{sw} + P_{tr} + P_{L_k} \qquad (1)$$

Table 1. Operating characteristics of the DAB converter

Parameter	Value	Unit
Primary side dc-link voltage, v_{dc1}	400	V
Secondary side dc-link voltage, v_{dc2}	200	V
Nominal operating power	3600	W
Switching frequency, f_{sw}	100	kHz
Sampling frequency, f_s	50	kHz
Transformation ratio, n	2:1	–
Leakage inductance (primary), L_k	3.5	μH
Primary side dc-link capacitor, C_1	1680	μF
Secondary side dc-link capacitor, C_2	1680	μF
Secondary side resistive load	11.11	Ω

3   Simulation Results of Each Modulation Technique

As mentioned, each modulation technique will be studied and validated in this section through the implementation of computer simulations, comparing voltage and current waveforms in the most diverse points of the DAB converter. On the other hand, the triggering gate signals of the semiconductors S_1, S_3, S_5 e S_7 are shown, essential to the detailed study proposed in this paper.

3.1   DAB Operating as a Single Active Bridge (SAB) Converter Under Duty-Cycle Modulation

With the implementation of this modulation technique, it is intended to demonstrate the behavior of the system presuming the choice of a SAB topology, in which one of the H-bridges enables switching under the duty-cycle modulation, whilst the other assumes a passive behavior since it will be the semiconductor's free-wheeling diodes who provides a path for the current. Such architecture facilitates the implementation of this modulation technique and provides greater robustness to the system. Thus, in order to regulate the voltage on the secondary dc-link (v_{dc2}) at 200 V (previously established reference value), a PI control algorithm was used, acting directly on the duty-cycle of the semiconductors S_1 and S_3, as can be seen in Fig. 2.

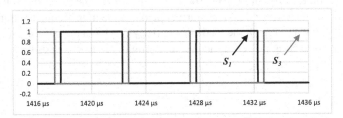

Fig. 2. Triggering gate signals of the semiconductors S_1 and S_3 using duty-cycle modulation.

Therefore, the current and voltage waveforms in the high-frequency transformer were monitored during a moment that v_{dc2} is very close to its steady-state reference value (200 V). As shown in Fig. 3 (a), the currents in the primary and secondary side of the transformer (i_{pri} and i_{sec}), despite having, respectively, peaks of 17.51 A and 35.02 A, present a null mean value and are in phase with the voltage waveforms at the terminals of the transformer (v_{pri} and v_{sec}). Moreover, it is verified that i_{pri} always takes half the instantaneous value of i_{sec}, thus respecting the applied transformation ratio. On the other hand, v_{pri} and v_{sec} can be observed in Fig. 3 (b), and, through their analysis, it is verified that v_{sec} has a maximum value close to 200 V, thus proving the veracity of the applied control algorithm.

Since v_{sec} is representative of the reflection of v_{pri} on the secondary side, the transfer of energy in the high-frequency transformer is slower and more limited, a direct consequence of the gradual, and also slower, increase of i_{pri}. In this way, it is possible to conclude that the energy transfer range is quite short and time consuming. Through the

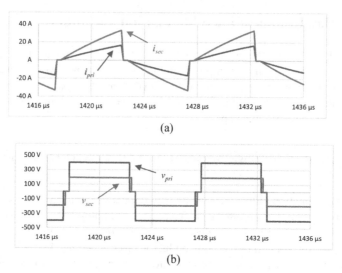

(a)

(b)

Fig. 3. Waveforms in the primary and secondary side of the high-frequency transformer using duty-cycle modulation: (a) Currents; (b) Voltages.

analysis of Fig. 3 (b), it is possible to verify a small mismatch between v_{pri} and v_{sec}, caused by the voltage drop in L_k. The lower the value of L_k, the lower the voltage drop, the aforementioned mismatch, and the current peaks in the high-frequency transformer.

Finally, as shown in Fig. 4, the load current (i_{load_sec}) has a mean value of 17.255 A, reason why the operating power of the system is slightly below the desired 3.6 kW.

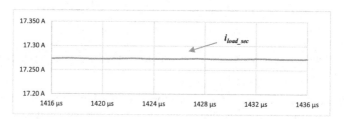

Fig. 4. Load current using duty-cycle modulation

3.2 Single Phase Shift (SPS)

The phase shift modulation techniques, compared to the aforementioned duty-cycle modulation, present a large number of advantages, highlighting the efficiency provided to the isolated dc-dc converters and the fact that all semiconductor gate signals have a fixed duty-cycle of 50%. Regardless of the direction of the power, each of the eight semiconductors is always switching and, as mentioned, with a constant duty-cycle. However, the phenomenon that allows the power transfer to and from each side of the DAB converter is the mismatch between the voltages at the terminals of the high-frequency transformer. These two signals, when phase shifted, generate a voltage in L_k and, consequently, a certain current will flow through it (i_{L_k}). Depending on whether the phase shift is positive or negative, the direction of i_{L_k} will be changed and the power will flow in accordance. In other words, if v_{pri} is ahead of v_{sec}, the power will flow from the primary side to the secondary one. If the opposite happens, power will flow from the secondary to the primary. It will be the adjustment of the phase mismatch between v_{pri} and v_{sec} that will regulate the value of transferred power: the higher its value, the greater the transferred power. However, as aforementioned, for simplicity reasons, this paper only analyzes the power flow from the primary to the secondary side.

In systems in which the value of v_{dc1} is very close to v_{dc2}, SPS modulation is usually used. However, the great limitation of this technique is the existence of only one degree of freedom, i.e., the only variable that can be controlled is the phase angle between each of the H-bridges (outer phase angle, D_0). This fact leads to higher reactive power values and circulating currents in the DAB, especially in times of non-correspondence in the voltage values on each side ($v_{dc1} \neq n v_{dc2}$). Thus, the converter losses tend to be naturally higher, as well as the current peaks in the semiconductors.

For this particular application case, the value of D_0 will be generated using a PI algorithm, resulting from the comparison between the reference of 200 V and the measured value of v_{dc2}. Since only degree of freedom is considered, the waveform of v_{pri} and v_{sec} presents two voltage levels, as can be seen in Fig. 5 (a). Looking at this figure, it is possible to verify that v_{sec} is delayed in relation to v_{pri}, thus proving the energy transfer from the primary side to the secondary one. This phase shift represents the value generated by the PI algorithm and can be seen in greater detail in Fig. 5 (b).

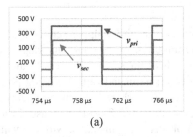
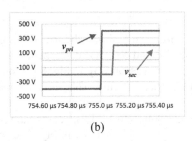

(a)   (b)

Fig. 5. Voltage waveforms in the primary and secondary side of the high-frequency transformer using SPS modulation: (a) During a complete cycle; (b) In detail.

In Fig. 6, on the other hand, the current waveforms on each side of the high-frequency transformer are shown. As in the duty-cycle modulation technique (operating as SAB), these have null mean value and are in accordance with the v_{pri} and v_{sec} waveforms, i.e., the currents have a positive value when the voltage is also positive, the same happening for the situation when the current is negative.

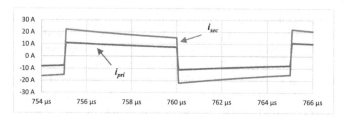

Fig. 6. Current waveforms in the primary and secondary side of the high-frequency transformer using SPS modulation.

Figure 7 shows the voltage and current waveforms in the resistive load connected in parallel with the secondary dc-link. As it can be seen, in steady-state, the voltage assumes an average value of 199.03 V while the current is at 17.915 A. Thus, the active power in the load has an approximate value of 3.566 kW, a value considered very close to the desired one (3.6 kW).

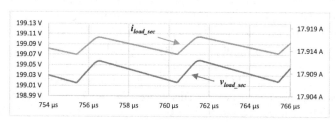

Fig. 7. Voltage and current waveforms in the resistive load in steady-state using SPS modulation.

In turn, Fig. 8 shows the triggering gate signals of each leg upper semiconductor (S_1, S_3, S_5, and S_7). As mentioned, the SPS technique only considers one degree of freedom, the reason why in Fig. 8 (a) the gate signals of the semiconductors S_1 and S_3 are 180° phase shifted. This same degree of freedom is related to the existing phase shift between the two H-bridges (D_0), visible in Fig. 8 (b), where S_5 assumes a delay of 2.88° in relation to S_1. Lastly, since the gate signal applied to S_7 is 180° phase shifted from the one applied to S_5, in Fig. 8 (c) the sum of this 180° mismatch with the value of D_0 is shown, thus making a total phase lag of 182.88° between S_1 and S_7. It should also be noted that, as expected, the PWM signals applied to S_2, S_4, S_6, and S_8 will be complementary to the ones showed in Fig. 8.

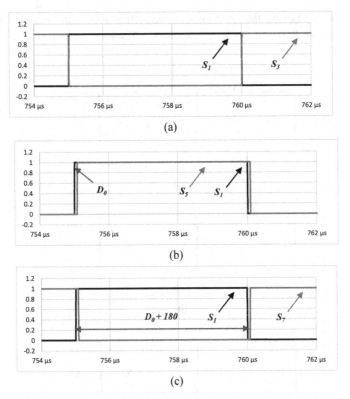

Fig. 8. Triggering gate signals using SPS modulation: (a) S_1, S_3; (b) S_1, S_5; (c) S_1, S_7.

3.3 Dual Phase Shift (DPS)

The DPS modulation, compared to SPS, contemplates a new degree of freedom, thus making a total of two. These degrees of freedom are related to the phase shift between the two H-bridges (D_0, as for the SPS) and the legs of each H-bridge (inner phase angle, D_1). With the addition of D_1, the efficiency of the DAB is improved, since the value of reactive power, circulating currents, and current peaks will be reduced. Although the ZVS operating range is not considered in these simulations, in theory, it could also be expanded. On the other hand, the inclusion of D_1 will also allow obtaining a three level voltage waveform at the terminals of the high-frequency transformer, but to really optimize energy efficiency levels, the values of D_0 and D_1 must be calculated based on duty cycle and phase shift modulations. Thus, as in the SPS technique, the value of D_0 will be obtained using a PI algorithm, whereas, D_1 will be calculated with the aid of (2) and (3). This equation will vary the value of D_1 according to the relation between v_{dc1} and v_{dc2}, being as much greater as the value of the ratio.

The values of D_0 and D_1 can be observed in Fig. 9 (a) and Fig. 9 (b), respectively. In the image presented below, the triggering gate signals of S_1, S_3, S_5, and S_7 using DPS modulation are shown. By analyzing Fig. 9, it is observed that the signals applied to S_3,

S_5, and S_7 are ahead of S_1, thus proving the energy flow direction. In turn, Fig. 9 (c) a comparison is made between the signals applied to S_1 and S_7, lagged $D_0 + D_1$ degrees.

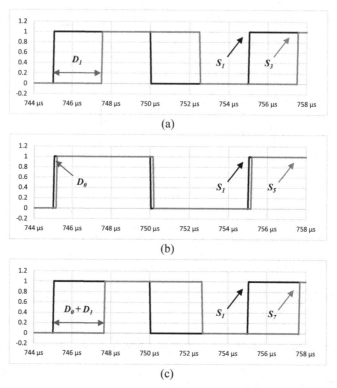

Fig. 9. Triggering gate signals using DPS modulation: (a) S_1, S_3; (b) S_1, S_5; (c) S_1, S_7.

$$error = \left| n - \frac{v_{dc1}}{v_{dc2}} \right| \qquad (2)$$

$$D_1 = 180\, e^{-2.197\, error} \qquad (3)$$

As mentioned, with the addition of a new degree of freedom, a voltage waveform with three levels will be generated at the terminals of the high-frequency transformer. Given the higher number of levels, the efficiency of this magnetic element will also be higher and the THD significantly lower, which is why multilevel topologies have received special prominence in power electronics systems to mitigate power quality problems. The waveforms of v_{pri} and v_{sec} are presented in Fig. 10, being possible to clearly observe the existing phase lag between them. Thus, with v_{pri} ahead of v_{sec}, it is proven, once again, the direction of the energy flow.

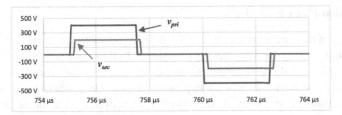

Fig. 10. Voltage waveforms in the primary and secondary side of the high-frequency transformer using DPS modulation.

In addition, in Fig. 11 are shown the current waveforms in the high-frequency transformer with the employment of a DPS modulation. By analyzing this graphic, it is concluded that, once again, the mean value of these currents is null, even if the latter does not assume zero value when v_{pri} and v_{sec} does. Although the current peak is higher than in SPS modulation, the stress will be much lower.

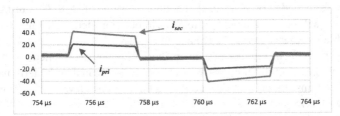

Fig. 11. Current waveforms it the primary and secondary side of the high-frequency transformer using DPS modulation.

However, it is also necessary to monitor and regulate the voltage on the secondary dc-link, being essential to maintaining its level close to the 200 V reference. In Fig. 12 it is shown the regulation of v_{dc2} at a transient state, in which $t = 0.63$ ms is representative of the moment that the switching signals of the secondary bridge are enabled. With the adoption of this measure, a pre-charge of the dc-link capacitors is carried out, thus avoiding overcurrent, reducing total losses, and guaranteeing the integrity of the hardware.

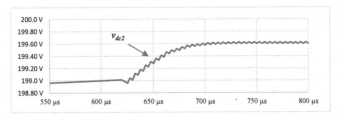

Fig. 12. Voltage waveform in the secondary dc-link during a transient state using DPS modulation.

3.4   Extended Phase Shift (EPS)

As mentioned in the introduction of this paper, the EPS modulation is normally employed in medium/high voltage power electronics systems. In this context, when compared to the SPS technique, EPS modulation is capable of reducing the values of circulating current and conduction losses when a large amount of power needs to be transferred to and from each side of an isolated dc-dc converter. Moreover, this technique enables and improves the soft-switching capacity of the converter and presents a large ZVS operating range.

However, in low power systems, as is the case, the EPS modulation has a reduced impact and may even provoke worse results than SPS. Under light load conditions, the operating range decreases, the reason why the obtained simulation results, compared to SPS modulation, show higher current peaks and semiconductor stress.

This modulation technique, as for DPS, considers two degrees of freedom, once again, related to the phase shift between v_{pri} and v_{sec} and between the legs of the primary H-bridge. In Fig. 13 (a), the influence D_1 is shown, being possible to observe the phase lag between S_1 and S_3. However, the difference between DPS and EPS lies in the signal applied to the semiconductor S_7: instead of adding the value of D_0 to D_1, during the EPS modulation, S_7 will be complementary to S_5, without considering the influence of D_1, as it is possible to observe in Fig. 13 (b) and in Fig. 13 (c).

As it can be seen in Fig. 14 (a), given the PWM signals applied to each semiconductor, voltages with different waveforms will be generated at the terminals of the high-frequency transformer, with v_{pri} presenting three voltage levels (v_{dc1}, 0 and $-v_{dc1}$) and v_{sec} only two (v_{dc2} and $-v_{dc2}$).

Nonetheless, during the moments that v_{pri} and v_{sec} assume their respective absolute maximum values (400 V and 200 V, represented by t_1, t_2, and t_3 in the figures presented below), i_{pri} and i_{sec} increase more sharply, showing peak values that are not compatible with this application case (Fig. 14 (b)). A possible solution would be to increase the value of D_1 or decrease D_0, however, the regulation of v_{dc2} to the reference value of 200 V would be impossible to achieve. In turn, in the event of a possible change in the DAB converter parameters, increasing the value of L_k or the system's f_s, the current peaks of i_{pri} and i_{sec} would be considerably lower.

Lastly, and once again, the regulation of v_{dc2} can be observed in Fig. 15 during a transient state, coincidental with the moment when the capacitor pre-charge is finished. As it is possible to contemplate, the steady-state value of v_{dc2} is 202.7 V, thus resulting in a difference of 2.7 V for the pre-established reference.

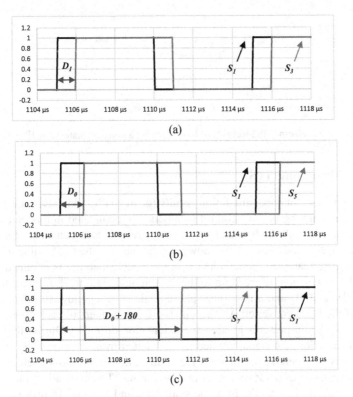

Fig. 13. Triggering gate signals using EPS modulation: (a) S_1, S_3; (b) S_1, S_5; (c) S_1, S_7.

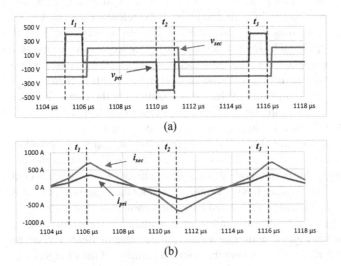

Fig. 14. Waveforms in the primary and secondary side of the high-frequency transformer using EPS modulation: (a) Voltages; (b) Currents.

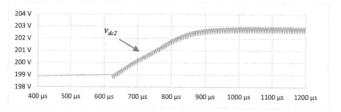

Fig. 15. Voltage waveform in the secondary dc-link during a transient state using EPS modulation.

3.5 Triple Phase Shift (TPS)

In this section, the obtained simulation results for the TPS modulation are analyzed. In the literature, among the five control techniques contemplated in this paper, TPS is considered the one with the best energy efficiency results, being capable of mitigating circulating currents and high reactive power values, change the RMS value of i_{L_k}, and, consequently, decrease the semiconductor stress. Such capability is obtained with the addition of a new degree of freedom to the system, making a total of three: the afore-mentioned D_0 and D_1 and a new inner phase angle (D_2), employed to the secondary H-bridge of the DAB converter. All the angles are independent of each other, which is why TPS modulation presents greater implementation difficulty. During the adoption of a SPS modulation, the signal applied to S_7 resulted from the sum of D_0 and D_1, a measure that is considered a limitation to the optimization of the soft-switching capacity and to the expansion of the ZVS range. In turn, during TPS modulation, the triggering gate signal applied to S_7 results from the sum of D_0 and D_2 (Fig. 16 (c)), i.e., the new degree of freedom here contemplated. The analysis of Fig. 16 (a) and Fig. 16 (b) shows that the triggering gate signals applied to S_3 and S_5 are similar to EPS and DPS cases, adding the value of D_1 and D_0, respectively, to the reference assigned to S_1.

Through the realization of computer simulations, it was concluded that the TPS technique, due to the inclusion of a new degree of freedom, is a very flexible modulation, capable of being applied in low power systems, which was not the case for EPS. As Fig. 17 (a) shows, with the adjustment of the value of D_2 it is possible to change the time that v_{sec} assumes its absolute maximum value (200 V), allowing the energy transfer at a higher or reduced rate, depending on the application case. That fact can be confirmed by the analysis of Fig. 17 (b), in which i_{pri} and i_{sec} are presented.

It should also be noted that for the purpose of analyzing the behavior of the TPS modulation in the face of oscillations in the value of D_2, during the computational simulations the value of this angle was varied manually, i.e., without any control strategy applied. Nonetheless, when it is important to consider the expansion of the ZVS operating range, this value should be calculated based on appropriate strategies.

As was the case for the aforementioned modulation techniques, the regulation of v_{dc2} is verified in Fig. 18, assuming a steady-state value of 199.8 V (extremely close to its reference, 200 V), thus proving the correct operation of the PI algorithm.

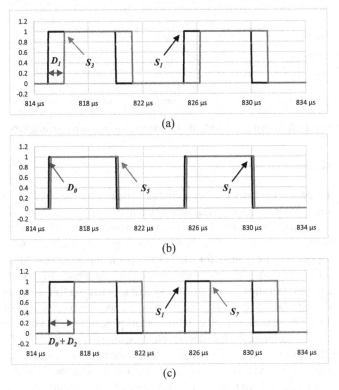

Fig. 16. Triggering gate signals using TPS modulation: (a) S_1, S_3; (b) S_1, S_5; (c) S_1, S_7.

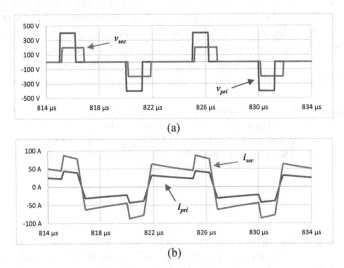

Fig. 17. Waveforms in the primary and secondary side of the high-frequency transformer using TPS modulation: (a) Voltages; (b) Currents.

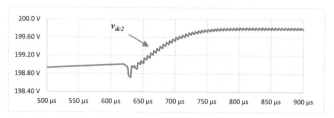

Fig. 18. Voltage waveform in the secondary dc-link during a transient state using TPS modulation.

4 Conclusions

Throughout the paper, five modulation techniques that could be applied to control a dual active bridge (DAB) converter were studied, namely duty-cycle modulation, single phase shift (SPS), dual phase shift (DPS), extended phase shift (EPS), and triple phase shift (TPS). The results obtained in computer simulations were later compared, showing the different triggering gate signals for each modulation and analyzing the current and voltage waveforms in the most diverse points of the DAB converter. In turn, these results were obtained for an operating power of 3.6 kW and a switching frequency of 100 kHz. Taking into account the assigned simulation parameters, it was possible to conclude that among all the aforementioned modulation techniques, only SPS, DPS, and TPS are suitable for this particular case. The duty-cycle modulation, when applied to a single active bridge (SAB) topology, presents great limitations during power transfer, since this process is carried out at a very low rate. On the other hand, the EPS modulation is more suitable for medium/high power electronics systems, being capable of transferring a certain power value in a short period of time, convenient for controlling high power dynamic systems with constant variations in its operating conditions. Lastly, among SPS, DPS and TPS modulation techniques, the last one is seen as the most flexible, since it enables three degrees of freedom (SPS presents one, and DPS two). However, its implementation difficulty is much higher and does not compensate the effort for this specific application case, being able to be applied in medium power systems that need to vary the RMS value of the current that flows through the leakage inductance.

Acknowledgments. This work has been supported by FCT – Fundação para a Ciência e Tecnologia within the Project Scope: UIDB/00319/2020. This work has been supported by the FCT Project PV4SUSTAINABILITY Reference: 333203230 and by the project newERA4GRIDs PTDC/EEI-EEE/30283/2017. Tiago Sousa is supported by the doctoral scholarship SFRH/BD/134353/2017 granted by FCT.

References

1. de Doncker, R.W.A.A., Divan, D.M., Kheraluwala, M.H.: A three-phase soft-switched high-power-density DC/DC converter for high-power applications. IEEE Trans. Ind. Appl. **27**(1), 63–73 (1991). https://doi.org/10.1109/28.67533
2. Mueller, J.A., Kimball, J.W.: Modeling dual active bridge converters in DC distribution systems. IEEE Trans. Power Electron. **34**(6), 5867–5879 (2019). https://doi.org/10.1109/TPEL.2018.2867434

3. Dao, N.D., Lee, D.C., Phan, Q.D.: High-efficiency SiC-based isolated three-port DC/DC converters for hybrid charging stations. IEEE Trans. Power Electron. **35**(10), 10455–10465 (2020). https://doi.org/10.1109/TPEL.2020.2975124

4. Wang, D., Nahid-Mobarakeh, B., Emadi, A.: Second harmonic current reduction for a battery-driven grid interface with three-phase dual active bridge DC-DC converter. IEEE Trans. Industr. Electron. **66**(11), 9056–9064 (2019). https://doi.org/10.1109/TIE.2019.2899563

5. Sun, Y., Gao, Z., Fu, C., Wu, C., Chen, Z.: A hybrid modular DC solid-state transformer combining high efficiency and control flexibility. IEEE Trans. Power Electron. **35**(4), 3434–3449 (2020). https://doi.org/10.1109/TPEL.2019.2935029

6. Liu, J., Yang, J., Zhang, J., Nan, Z., Zheng, Q.: Voltage balance control based on dual active bridge DC/DC converters in a power electronic traction transformer. IEEE Trans. Power Electron. **33**(2), 1696–1714 (2018). https://doi.org/10.1109/TPEL.2017.2679489

7. Kwak, B., Kim, M., Kim, J.: Inrush current reduction technology of DAB converter for low-voltage battery systems and DC bus connections in DC microgrids. IET Power Electron. **13**(8), 1528–1536 (2020). https://doi.org/10.1049/iet-pel.2019.0506

8. Hu, J., Joebges, P., Pasupuleti, G.C., Averous, N.R., de Doncker, R.W.: A maximum-output-power-point-tracking-controlled dual-active bridge converter for photovoltaic energy integration into MVDC grids. IEEE Trans. Energy Convers. **34**(1), 170–180 (2019). https://doi.org/10.1109/TEC.2018.2874936

9. Chakraborty, S., Chattopadhyay, S.: Fully ZVS, minimum RMS current operation of the dual-active half-bridge converter using closed-loop three-degree-of-freedom control. IEEE Trans. Power Electron. **33**(12), 10188–10199 (2018). https://doi.org/10.1109/TPEL.2018.2811640

10. Roggia, L., Costa, P.F.S.: Comparative analysis between integrated full-bridge-forward and dual active bridge DC–DC converters. Electron. Lett. **54**(4), 231–233 (2018). https://doi.org/10.1049/el.2017.3326

11. Liu, P., Chen, C., Duan, S.: An optimized modulation strategy for the three-level DAB converter with five control degrees of freedom. IEEE Trans. Industr. Electron. **67**(1), 254–264 (2020). https://doi.org/10.1109/TIE.2019.2896209

12. Xuan, Y., Yang, X., Chen, W., Liu, T., Hao, X.: A novel NPC dual-active-bridge converter with blocking capacitor for energy storage system. IEEE Trans. Power Electron. **34**(11), 10635–10649 (2019). https://doi.org/10.1109/TPEL.2019.2898454

13. Chan, Y.P., Yaqoob, M., Wong, C.S., Loo, K.H.: Realization of high-efficiency dual-active-bridge converter with reconfigurable multilevel modulation scheme. IEEE J. Emerg. Sel. Top. Power Electron. **8**(2), 1178–1192 (2020). https://doi.org/10.1109/JESTPE.2019.2926070

14. Wu, J., Li, Y., Sun, X., Liu, F.: A new dual-bridge series resonant DC-DC converter with dual tank. IEEE Trans. Power Electron. **33**(5), 3884–3897 (2018). https://doi.org/10.1109/TPEL.2017.2723640

15. Rolak, M., Sobol, C., Malinowski, M., Stynski, S.: Efficiency optimization of two dual active bridge converters operating in parallel. IEEE Trans. Power Electron. **35**(6), 6523–6532 (2020). https://doi.org/10.1109/TPEL.2019.2951833

16. Jeung, Y.C., Lee, D.C.: Voltage and current regulations of bidirectional isolated dual-active-bridge DC-DC converters based on a double-integral sliding mode control. IEEE Trans. Power Electron. **34**(7), 6937–6946 (2019). https://doi.org/10.1109/TPEL.2018.2873834

17. Li, X., Wu, F., Yang, G., Liu, H., Meng, T.: Dual-period-decoupled space vector phase-shifted modulation for DAB-based three-phase single-stage AC-DC converter. IEEE Trans. Power Electron. **35**(6), 6447–6457 (2020). https://doi.org/10.1109/TPEL.2019.2950059

18. Takagi, K., Fujita, H.: Dynamic control and performance of a dual-active-bridge DC-DC converter. IEEE Trans. Power Electron. **33**(9), 7858–7866 (2018). https://doi.org/10.1109/TPEL.2017.2773267

19. Xia, P., Shi, H., Wen, H., Bu, Q., Hu, Y., Yang, Y.: Robust LMI-LQR control for dual-active-bridge DC-DC converters with high parameter uncertainties. IEEE Trans. Transp. Electrif. **6**(1), 131–145 (2020). https://doi.org/10.1109/TTE.2020.2975313

20. Shi, H., Wen, H., Chen, J., Hu, Y., Jiang, L., Chen, G.: Minimum-reactive-power scheme of dual-active-bridge DC-DC converter with three-level modulated phase-shift control. IEEE Trans. Ind. Appl. **53**(6), 5573–5586 (2017). https://doi.org/10.1109/TIA.2017.2729417

21. Karthikeyan, V., Gupta, R.: FRS-DAB converter for elimination of circulation power flow at input and output ends. IEEE Trans. Industr. Electron. **65**(3), 2135–2144 (2018). https://doi.org/10.1109/TIE.2017.2740853

22. Vazquez, N., Liserre, M.: Peak current control and feed-forward compensation of a DAB converter. IEEE Trans. Industr. Electron. **67**(10), 8381–8391 (2020). https://doi.org/10.1109/TIE.2019.2949523

23. Hebala, O.M., Aboushady, A.A., Ahmed, K.H., Abdelsalam, I.: Generic closed-loop controller for power regulation in dual active bridge DC-DC converter with current stress minimization. IEEE Trans. Industr. Electron. **66**(6), 4468–4478 (2019). https://doi.org/10.1109/TIE.2018.2860535

24. Yaqoob, M., Loo, K.H., Lai, Y.M.: A four-degrees-of-freedom modulation strategy for dual-active-bridge series-resonant converter designed for total loss minimization. IEEE Trans. Power Electron. **34**(2), 1065–1081 (2019). https://doi.org/10.1109/TPEL.2018.2865969

25. Liu, P., Duan, S.: A hybrid modulation strategy providing lower inductor current for the DAB converter with the aid of DC blocking capacitors. IEEE Trans. Power Electron. **35**(4), 4309–4320 (2020). https://doi.org/10.1109/TPEL.2019.2937161

26. Qin, Z., Shen, Y., Loh, P.C., Wang, H., Blaabjerg, F.: A dual active bridge converter with an extended high-efficiency range by DC blocking capacitor voltage control. IEEE Trans. Power Electron. **33**(7), 5949–5966 (2018). https://doi.org/10.1109/TPEL.2017.2746518

27. Xu, G., Sha, D., Xu, Y., Liao, X.: Hybrid-bridge-based DAB converter with voltage match control for wide voltage conversion gain application. IEEE Trans. Power Electron. **33**(2), 1378–1388 (2018). https://doi.org/10.1109/TPEL.2017.2678524

28. Xiao, Y., Zhang, Z., Andersen, M.A.E., Sun, K.: Impact on ZVS operation by splitting inductance to both sides of transformer for 1-MHz GaN based DAB converter. IEEE Trans. Power Electron. **35**(11), 11988–12002 (2020). https://doi.org/10.1109/TPEL.2020.2988638

29. Dung, N.A., Chiu, H.J., Lin, J.Y., Hsieh, Y.C., Liu, Y.C.: Efficiency optimisation of ZVS isolated bidirectional DAB converters. IET Power Electron. **11**(8), 1–8 (2018). https://doi.org/10.1049/iet-pel.2017.0723

30. Garcia-Bediaga, A., Villar, I., Rujas, A., Mir, L.: DAB modulation schema with extended ZVS region for applications with wide input/output voltage. IET Power Electron. **11**(13), 1–8 (2018). https://doi.org/10.1049/iet-pel.2018.5332

31. Xu, G., Sha, D., Xu, Y., Liao, X.: Dual-transformer-based DAB converter with wide ZVS range for wide voltage conversion gain application. IEEE Trans. Industr. Electron. **65**(4), 3306–3316 (2018). https://doi.org/10.1109/TIE.2017.2756601

32. Shi, H., et al.: Minimum-backflow-power scheme of DAB-based solid-state transformer with extended-phase-shift control. IEEE Trans. Ind. Appl. **54**(4), 3483–3496 (2018). https://doi.org/10.1109/TIA.2018.2819120

33. Shen, K., et al.: ZVS control strategy of dual active bridge DC/DC converter with triple-phase-shift modulation considering RMS current optimization. J. Eng. **2019**(18), 4708–4712 (2019). https://doi.org/10.1049/joe.2018.9341

34. Calderon, C., et al.: General analysis of switching modes in a dual active bridge with triple phase shift modulation. Energies **11**(9), 2419 (2018). https://doi.org/10.3390/en11092419

35. Bu, Q., Wen, H., Wen, J., Hu, Y., Du, Y.: Transient DC bias elimination of dual-active-bridge DC-DC converter with improved triple-phase-shift control. IEEE Trans. Industr. Electron. **67**(10), 8587–8598 (2020). https://doi.org/10.1109/TIE.2019.2947809

36. Dai, T., et al.: Research on transient DC bias analysis and suppression in EPS DAB DC-DC converter. IEEE Access **8**, 61421–61432 (2020). https://doi.org/10.1109/ACCESS.2020.298 3090

37. Wu, F., Feng, F., Gooi, H.B.: Cooperative triple-phase-shift control for isolated DAB DC-DC converter to improve current characteristics. IEEE Trans. Industr. Electron. **66**(9), 7022–7031 (2019). https://doi.org/10.1109/TIE.2018.2877115

38. Luo, S., Wu, F., Wang, G.: Improved TPS control for DAB DC-DC converter to eliminate dual-side flow back currents. IET Power Electron. **13**(1), 32–39 (2020). https://doi.org/10.1049/iet-pel.2019.0562

39. Liu, X., et al.: Novel dual-phase-shift control with bidirectional inner phase shifts for a dual-active-bridge converter having low surge current and stable power control. IEEE Trans. Power Electron. **32**(5), 4095–4106 (2017). https://doi.org/10.1109/TPEL.2016.2593939

40. Hou, N., Song, W., Li, Y., Zhu, Y., Zhu, Y.: A comprehensive optimization control of dual-active-bridge DC-DC converters based on unified-phase-shift and power-balancing scheme. IEEE Trans. Power Electron. **34**(1), 826–839 (2018). https://doi.org/10.1109/TPEL.2018.281 3995

41. Fritz, N., Rashed, M., Bozhko, S., Cuomo, F., Wheeler, P.: Flux control modulation for the dual active bridge DC/DC converter. J. Eng. **2019**(17), 4353–4358 (2019). https://doi.org/10.1049/joe.2018.8014

42. Zengin, S., Boztepe, M.: A novel current modulation method to eliminate low-frequency harmonics in single-stage dual active bridge AC-DC converter. IEEE Trans. Industr. Electron. **67**(2), 1048–1058 (2020). https://doi.org/10.1109/TIE.2019.2898597

43. Kumar, A., Bhat, A.H., Agarwal, P.: Comparative analysis of dual active bridge isolated DC to DC converter with single phase shift and extended phase shift control techniques. In: 2017 6th International Conference on Computer Applications in Electrical Engineering - Recent Advances, CERA 2017, vol. 2018-January, pp. 397–402 (2018). https://doi.org/10.1109/CERA.2017.8343363

44. Kumar, B.M., Kumar, A., Bhat, A.H., Agarwal, P.: Comparative study of dual active bridge isolated DC to DC converter with single phase shift and dual phase shift control techniques. In: 2017 Recent Developments in Control, Automation and Power Engineering, RDCAPE 2017, vol. 3, pp. 453–458 (2018). https://doi.org/10.1109/RDCAPE.2017.8358314

45. Kumar, A., Bhat, A.H., Agarwal, P.: Comparative analysis of dual active bridge isolated DC to DC converter with double phase shift and triple phase shift control techniques. In: 2017 Recent Developments in Control, Automation and Power Engineering, RDCAPE 2017, vol. 3, pp. 257–262 (2017). https://doi.org/10.1109/RDCAPE.2017.8358278

46. Kayaalp, I., Demirdelen, T., Koroglu, T., Cuma, M.U., Bayindir, K.C., Tumay, M.: Comparison of different phase-shift control methods at isolated bidirectional DC-DC converter. Int. J. Appl. Math. Electron. Comput. **4**(3), 68 (2016). https://doi.org/10.18100/ijamec.60506

Demand Response; Energy; Smart Homes

Smart Greenhouse Project: A Sustainability-Focused Learning Experience for Undergraduates

Rick Todd Kaske Jr.[✉], Brady Connaher, and Mohammad Upal Mahfuz

University of Wisconsin-Green Bay, Green Bay, WI 54311, USA
{kaskrt28,connba24,mahfuzm}@uwgb.edu

Abstract. In this paper, a smart greenhouse project has been presented as an example of a sustainability-focused learning experience for undergraduates. While project-based instruction and learning experience has already gained enough momentum in several fields of education, the inclusion of sustainability perspectives from the science and engineering fields of studies is yet to make its room in the current course curricula as well as in the mindset of new generations of engineering undergraduates. This smart greenhouse project is an example of how sustainability can be brought into a classroom setting of engineering and technology programs as a project-based learning experience. The objective of this smart greenhouse project is to create an automated system capable of growing vegetation with little human input by utilizing electricity, computer programming, and a microcontroller operation. While this project was implemented by a group of two students with electrical engineering technology (EET) major as parts of their *Introduction to Programming* (ET 142) and *Supervisory Control and Data Acquisition* (ET 342) courses requirements at the University of Wisconsin-Green Bay, USA, a similar sustainability-focused learning approach can also be applied successfully in other courses at different levels of engineering and technology programs at other academic institutions.

Keywords: Project-based learning · Smart greenhouse · Sustainability · Sustainable design

1 Introduction

Sustainability and sustainable development have been the topics of discussions in recent years among researchers of all disciplines including education [1, 2]. In the field of science, technology, engineering, and mathematics (STEM), the opportunities of creating new devices, practices, and algorithms have made it even more important to consider sustainability matters for long-term future [1] and the new generations of scientists, engineers, and technologists. In STEM fields, it has become increasingly important that

---

B. Connaher—Both first and second authors contributed equally to this work.

© ICST Institute for Computer Sciences, Social Informatics and Telecommunications Engineering 2021
Published by Springer Nature Switzerland AG 2021. All Rights Reserved
J. L. Afonso et al. (Eds.): SESC 2020, LNICST 375, pp. 53–62, 2021.
https://doi.org/10.1007/978-3-030-73585-2_4

the students get hands-on experience to build things in order to meet expectations of their future employers. As a result, it has now been evident that sustainability must be addressed in the hands-on experience that students receive at academic institutions especially in their undergraduate level before they enter actual professional careers [3, 4].

Project-based learning experience is now-a-days quite effective and thus popular in STEM fields [5]. Project-based learning is found to increase students' interests in the STEM fields mainly because students learn more when they can build things with the knowledge that they receive in classrooms [6]. Projects also encourage students to solve real-life problems and work with a team. Therefore, in order to be successful professionals in the future, project-based learning has now been used in many STEM fields [7, 8].

Traditionally, in a STEM undergraduate course curriculum, a typical computer programming class has a theory part, where students learn fundamental concepts of programming, accompanied by a laboratory exercises part, where students apply those concepts that they learn in the theory part to do actual programming solutions. The laboratory exercises part clearly provides an opportunity for students to learn hands-on programming practices. On the other hand, in the current times of industrial automation and controls, the topic of *Supervisory Control and Data Acquisition* (SCADA) is extremely important for a successful automation system that is spread over a wide geographical area where an operator in a central hub, e.g. a master station, could see all of the system components in the user interface and send control (e.g. supervisory command) signals to the field devices. Contents of a SCADA course typically include a theoretical part discussing the various aspects of data acquisition and supervisory controls and a laboratory part providing students with an opportunity to get hands-on experience on SCADA software [9]. While in both the *Introduction to Programming* (ET 142) and *Supervisory Control and Data Acquisition* (ET 342) classes taught at the University of Wisconsin-Green Bay, USA, in order to allow students to receive an additional thorough hands-on experience of computer programming and SCADA, they are required to do a project work in each course where they apply computer programming and SCADA skills to provide real-life solutions. In addition, in order to address the emerging need of sustainable engineering practices and build future engineers with sustainability-focused mindset, students are encouraged to do their project on sustainability-focused engineering and technology solutions [2]. In this paper, one of these student projects on sustainability-focused engineering solutions in the smart cities aspects has been presented, which was performed by a group of two students in two parts as ET 142 and ET 342 courses requirements respectively at the University of Wisconsin-Green Bay, USA in the fall 2019 semester.

The rest of this paper is organized as follows: Sect. 2 details the background, implementation and experimental work, an explanation of functionality, and data collection and control systems used in this project. Section 3 presents an evaluation of the students' learning experience from this project followed by Sect. 4 reflecting on the instructor's experience on teaching this course. Finally, Sect. 5 provides some concluding remarks.

2   Smart Greenhouse Project

2.1   Motivation and Objective

The smart greenhouse project is an attempt to make indoor food production possible while also incorporating efficient resource utilization. The smart greenhouse is designed to provide an ideal environment for plant life to thrive with minimum human influence. This is accomplished by easily customizable programming as well as a scalable design. The objective of this project is to create an automated system capable of growing vegetation with little human input by utilizing electricity, programming, and a microcontroller operation. Scalable sustainable food production is one of the most important challenges facing a rapidly growing population. In order to meet this growing demand, the agriculture industry will have to modernize in order to increase efficiency and the success rate of crop production. The use of technology will also increase the ability of individuals to pursue self-sustainable food production in areas where vegetable production would otherwise be impossible. This is especially important in areas of the world where temperatures prevent or inhibit growing food year-round.

2.2   Implementation and Experimental Work

Most of the parts and components used in this project can be found in any standard electronics laboratory, electronics shop, or from any electronics vendor. The hardware and software used in this project are: Arduino Uno Microcontroller [10], Arduino Independent Development Environment (IDE), breadboard, DC motor, servo9g servo, DHT 11 temperature and humidity sensor, real time clock (RTC) module, relay module, UV light, power supply, wires, resistors, diode rectifier, and MOSFET.

Various experiments were performed on the project in order to achieve a satisfactory performance. An independent testing program has been developed in order to test and experiment with each individual module of the smart greenhouse project. In the current design of the project, it was discovered that, while the servo and the DC motor both work individually, they are unable to operate at the same time while utilizing the microcontroller's 5V power supply. It was experimented with isolating the servo on another circuit with another microcontroller but ultimately this became too convoluted from a programming aspect, as well as being an unnecessary cost increase. Instead, it was experimented with programming solutions. This required an adaptation of our smart greenhouse program to guarantee that the servo would not energize when the DC motor was functioning. The control circuit of the smart greenhouse project has been shown in Fig. 1. Different portions of the experimental setup have been shown in detail in Fig. 2 and Fig. 3.

2.3   Functional Description

The smart greenhouse is programmed via the Arduino IDE [10]. It utilizes a programming language that is a modified version of C++. The programming controls the Arduino Uno microcontroller. This program was prototyped on the Arduino Uno but with further modifications it could be converted to run on the cheaper, smaller, Arduino nano. There are three main inputs that the program operates on, namely, temperature, humidity, and time.

Temperature and humidity are provided via the DHT 11 temperature and humidity sensor. It is capable of converting the analog signals of temperature and humidity into digital values that the microcontroller can understand and display. The program utilizes these values to determine whether or not to vent the greenhouse. Venting is accomplished via a fan attached to a DC motor. It is important that the fan control program be easily customizable as different flora require different temperatures and different levels of humidity.

Date and time are provided via the RTC module. This module can be set to acquire its starting time from the PC it is connected to or it can be set as part of the first run of the system. Values from the RTC module are utilized for three purposes in the smart greenhouse program. The first output that utilizes time is the relay module. This allows a user to specify a time period for the UV light to be energized. This function again must be customizable not only for different flora but also for different growth stage needs. Watering is also controlled via the date and time. The user enters a time and duration in the program for watering to occur. The servo then opens the valve at the specified time and for the specified duration. The servo programming also allows customization for how open the valve is. A small degree of rotation allows for drip irrigation whereas a large degree of rotation allows for a much greater volume of water. The last function of the RTC module is to display time stamps for data collection.

A video link to the project with explanation and functionality is provided in [11].

2.4 Data Collection

The smart greenhouse currently collects data from various analog components. It collects time and date from the RTC module. It collects temperature and humidity from the DHT 11 temperature and humidity sensor. Component activation is also included in the programming. For example, when the fan is running, a message is displayed stating "Fan is on.". This is true for the water irrigation servo and the relay as well. This data is currently stored and displayed in the serial monitor built into the Arduino IDE. The program is set to send the data every five seconds. Each data communication includes a time stamp, the temperature of the unit, the humidity of the unit, and the status of various components. It can be scaled with an SD card reader module to facilitate long term storage of this data, which currently is not possible due to the serial monitors limited memory. This would also allow the data to be analyzed by utilizing Microsoft Excel in order to increase efficiency and production. Figure 4 shows the serial communication mimic diagram for the smart greenhouse project.

2.5 Control

Currently live control of the program has not been accomplished. In order to change any variables of the programming requires a change to the main program and then for the program to be reuploaded to the Arduino. Through further code work it would be able to provide control to the smart greenhouse through the serial monitor. This was accomplished partially with our testing program which allows user input to test each component. In an ideal system, it would have data collection and storage on an SD card for long term storage and utilize the serial monitor for immediate user review. This would then build in a set of commands that can be entered into the serial monitor which would allow for real time control of the smart greenhouse.

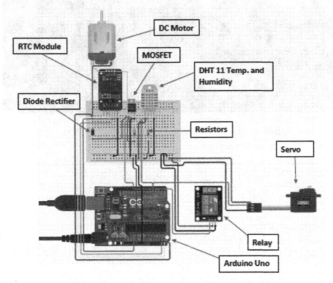

Fig. 1. The smart greenhouse project control circuit

Fig. 2. Smart greenhouse overview

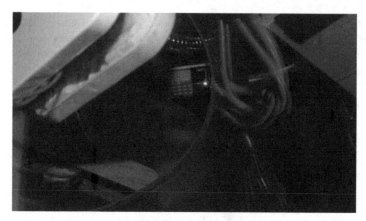

Fig. 3. DHT 11 temperature and humidity sensor

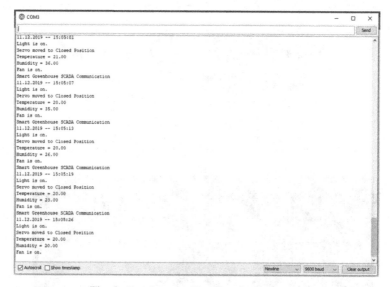

Fig. 4. Serial communication mimic diagram

3 Learning Experience Evaluation

To evaluate the students' learning experience, at the end of the project in the last week of the semester, the students were asked to provide their responses to a number of questions regarding their overall experience on this project. The questions were designed basically to evaluate the level of learning the students acknowledged in this project work. Finally, the students were asked if they would have any concluding remarks regarding this project. The following discussions in this section have been based on the responses from the students to these questions as well as any concluding remarks provided by the students.

Learning from this Project
Students responded positively to the project experience mentioning that the project increased not only their programming knowledge but also their knowledge of microcontrollers and circuits. The students specifically mentioned that the project allowed them to increase their ability to use programming knowledge to control electronic circuits. The students also added that they learned the importance of data collection accuracy in order to provide efficient control of a system. The students also responded that they learned that creating a control system could be challenging and in that they also mentioned about their system currently being controlled via programming changes but not meeting the requirement of a SCADA system providing easy methods of controls. Overall, student feedback on the questions suggests that they received the desired learning on programming knowledge from the project.

Solving any Technical Issues in this SCADA Project
When asked about solving any technical issues around the project study, the students responded constructively in that they often faced unexpected technical issues that they had to devote a significant level of efforts moving towards the solutions. The students acknowledged that technical troubleshooting is a major part of a successful completion of a project and a complete learning of the subject matter. The students mentioned that they had created and used a testing program in order to troubleshoot why they could not get all of their components to operate together. They also reported that they had encountered troubles while trying to initiate control of the smart greenhouse via the serial monitor, adding that while they were not able to solve this issue, they believed they were close to a programming solution that would allow real time serial monitor control. Overall, the students acknowledged that the solutions to technical issues should take into consideration a thorough investigation of the problem associated with all parts of the system.

Identifying the Importance of Learning Subject Matter
When asked about an important aspect that the project provided with in their learning of the subject matter (i.e., course), the students responded constructively in that the complexity of the programming aspect of the project reinforced the importance of a logical thought process when designing a product. They also added that, although initially the program seemed straight forward, they faced multiple new challenges as the program requirements grew based on practical needs.

Identifying a Real-Life, Practical, Engineering Problem
As a part of the evaluation, the students were asked whether they thought the project had given them an opportunity to investigate on a real-life, practical, engineering problem and explore its solutions with programming knowledge. Student responses to this question were positive answers where they mentioned about the practical engineering problem that they were dealing with and exploring solutions to. For instance, the students mentioned that innovation in food production is very much a real-life, practical, engineering problem and in that this smart greenhouse project had allowed them to investigate how people could produce food in a more localized and self-sustainable way. The students also added that the smart greenhouse, while not a final product, showed how food production can begin moving toward automation and increased self-sufficiency and served as a

functioning example of how programming could be used to control practical applications in real time. This clearly indicated that the students received the idea of the purpose of the project correctly and were able to apply programming knowledge towards solutions to real-life, practical, engineering problems.

Self-evaluation of the Project

The students were also asked to self-evaluate the project study. This means, they were asked whether they thought that the project helped them in learning the subject matter involved and if they were able to meet the learning outcomes of the course. The responses from the students suggested that the project helped them in learning the subject matter thoroughly. They acknowledged that the project helped them greatly in understanding the C++ programming language. They also mentioned that while the smart greenhouse was not identical to their initial vision, it came close to accomplishing everything they set out to accomplish. The students also mentioned that they were not able to facilitate long term storage of data collection or real-time control of the system; however, they mentioned that they believe that they were close and given enough time they believe that they would be able to integrate these two features into the system.

Addressing Sustainability

In the final question, the students were asked whether their SCADA project addressed sustainability. The objective of this question was to understand what and how the students thought about the concept of sustainability in their project and whether they were able to define sustainability of any form through their project work. Within the question, the students were also given some ideas on what the concept of sustainability could mean, for instance, managing environment well, reducing the burning of fossil fuel, cost-effectiveness (financial sustainability), or any other form of sustainability measures.

The responses from the students clearly suggest that the students were able to understand the concept of sustainability and then define it in the context of their project work. The students also responded with details in that their project idea and the project itself touched on several different areas of sustainability. The most obvious impact was personal sustainability by allowing an individual to supplement their food supply by growing some (or all) of their own foods. This also led to other forms of sustainability. The smart greenhouse can be optimized to use the minimum amount of water necessary to irrigate plants versus the largely wasteful watering that occurs in a farm field. More locally grown foods would also benefit the environment due to reducing the pollution that is produced by transporting crops long distances via various transportation vehicles. Growing own foods also helps a family become more financially sustainable as it is more cost-effective to grow foods than it is to buy.

4 Perspectives on Teaching Experience

This project was a part of the student's course requirement for the *Introduction to Programming* (ET 142) and *Supervisory Control and Data Acquisition* (ET 342) courses taught at the University of Wisconsin-Green Bay, USA. For these two courses, all students were required to complete a project part, as described in an earlier section, that involved computer programming and SCADA hardware and/or software approach respectively to solve a real-life problem. This smart greenhouse project idea came out of the students' curiosity on the subject matter and was not assigned to the students by the instructor. This made sure that the students did a reasonable amount of homework before proposing an idea for their projects, which let them understand that real-life engineering problems can be solved with computer programming and SCADA techniques. At the beginning of the semester, students were given a two weeks' time to come up with a project idea that would involve real-life problem solving. In the third week of the course, the students were required to discuss their project ideas with the instructor to make sure that the proposed project idea was doable within the semester's timeframe and that it is not unreasonably ambitious.

As first-time learners of computer programming and SCADA hardware and/or software, some students found the projects challenging; however, it was also found that, as the project moved forward, with teamwork and sufficient study and homework, the students gradually progressed towards successful implementation of the project work. Weekly discussions were held with the students to evaluate the progress of the project as well as any issues where instructor's help was needed. Students were asked to keep notes of the work done in a current week and any work that was going to be done in the next week. The instructor met all student groups weekly in order to get updates on the progress of their projects and if there was any question that the instructor could answer or any help needed. Project progress was tracked towards a successful completion as the semester gradually came to an end. It was found that the weekly progress meetings were useful for the students to keep in touch with the instructor to discuss about the project works effectively.

Overall, it was found that a project work, although proposed as a part of a particular course, does not necessarily relate to that course materials only. Eventually, the students studied some topics that were beyond the course syllabus and self-learned them for the sake of the project success. This made the students to investigate on topics and subject matters within and beyond the typical subject area of computer programming and SCADA and guided them to successful completion of the project. It was found that the investigative nature of the project paved a way to solidify their understanding of the required engineering subjects in relation to their project, which would definitely contribute to their future educational career.

5 Conclusion

In this paper, a smart greenhouse project has been presented with implementation details as an example of sustainability-focused learning experience for STEM undergraduates. We believe that the design of the smart greenhouse presented here would be useful in generating interests among students with further improvements possible in different parts of the design, especially in the directions of data collection and control strategies. Although the project was completed in the engineering school at the University of Wisconsin-Green Bay, the same style of sustainability-focused education can be offered at any other university and other programs different from engineering. As a programming course is a fundamental course for several disciplines within the STEM fields, this type of project-based approach could be considered as a means to experience sustainability-focused concepts for the new generations of undergraduates. The fundamentals on the subject matter of SCADA can be learned at the junior or senior year at an engineering undergraduate curriculum. Sustainability concepts in education, yet a significant challenge to incorporate in the current STEM curricula, are getting a momentum into education fields lately. The evaluation of learning experience as well as the perspectives on teaching experience should be useful to evaluate such a STEM course with an integrated sustainability-focused component, though content type might sometimes be a factor and require a case-by-case evaluation. However, it is strongly believed that such a project-based approach to sustainability focus on STEM curriculum would be useful in the long-term success in making the current students better citizens aware of sustainable resource management and development and thus meeting the United Nations sustainable development goals.

References

1. Wilson, D.: Exploring the intersection between engineering and sustainability education. Sustainability **11**(11), 3134 (2019)
2. Watson, M., Noyes, C., Rodgers, M.: Student perceptions of sustainability education in civil and environmental engineering at the Georgia Institute of Technology. J. Prof. Issues Eng. Educ. Pract. **139**(3), 235–243 (2013). https://doi.org/10.1061/(ASCE)EI.1943-5541.0000156
3. Natkin, L.W.: Education for sustainability: exploring teaching practices and perceptions of learning associated with a general education requirement. J. Gen. Educ. **65**(3–4), 216 (2016). https://doi.org/10.5325/jgeneeduc.65.3-4.0216
4. Chitchyan, R., et al.: Sustainability design in requirements engineering: state of practice, pp. 533–542 (2016)
5. Ferreira, L.C.B.C., Branquinho, O.C., Chaves, P.R., Cardieri, P., Fruett, F., Yacoub, M.D.: A PBL-based methodology for IoT teaching. IEEE Commun. Mag. **57**(11), 20–26 (2019). https://doi.org/10.1109/MCOM.001.1900242
6. Hosny, A., Gokhale, S.: Project-based learning in a Probabilistic Analysis course in the CS curriculum. In: 2017 IEEE Integrated STEM Education Conference (ISEC), 11–11 March 2017, pp. 24–29 (2017). https://doi.org/10.1109/ISECon.2017.7910243
7. Hosseinzadeh, N., Hesamzadeh, M.R.: Application of project-based learning (PBL) to the teaching of electrical power systems engineering. IEEE Trans. Educ. **55**(4), 495–501 (2012). https://doi.org/10.1109/TE.2012.2191588
8. Habash, R.W.Y., Suurtamm, C., Necsulescu, D.: Mechatronics learning studio: from "Play and Learn" to industry-inspired green energy applications. IEEE Trans. Educ. **54**(4), 667–674 (2011). https://doi.org/10.1109/TE.2011.2107555
9. Mahfuz, M.U.: Design and development of a SCADA course for engineering undergraduates. In: Proceedings of the 2020 IEEE Integrated STEM Education Conference (ISEC), 30 July–1 August 2020, NJ, USA, pp. 1–8 (2020)
10. Arduino Uno Microcontroller. https://store.arduino.cc/usa/arduino-uno-rev3. Accessed 4 Oct 2020
11. Video link to project. https://www.youtube.com/watch?v=x3rKZjWY7PE. Accessed 4 Oct 2020

Assessment of Renewable Energy Technologies Based on Multicriteria Decision Making Methods (MCDM): Ocean Energy Case

David A. Serrato$^{(\boxtimes)}$ ⓘ, Juan C. Castillo ⓘ, Laura Salazar ⓘ, Alejandro ⓘ, Harold Salazar, Juan Esteban Tibaquirá ⓘ, Álvaro Restrepo ⓘ, Juan Camilo ⓘ, and Tatiana Loaiza

Universidad Tecnológica de Pereira, Pereira, Colombia
daserrato@utp.edu.co

Abstract. Renewable energy technologies in OECD countries have been highly promoted for the purpose of producing cleaner energy and better life conditions for people in urban areas. However, developing countries require an additional extended analysis to assess the feasibility for their implementation, identify financial risks and settle emissions reduction. In Colombia, in the last several years, there have been more robust public policy strategies to expand alternative energy sources and accomplish the COP 21 limits of 20% GHG reduction before 2030. In this study, a methodology based on Multicriteria Decision Making Methods (MCDM), which is the result of a research project to assess comparatively new renewable energy technologies with renewable energy technologies currently used in the country based on technical, financial, and environmental criteria, has been developed. The methodology allows for the identification of the best and the worst alternatives from the output ranking, considering the numerical value of criteria placed on the decision matrix and the dominance index output. The methodology was tested to assess the comparison of ocean energy technologies in Colombia. Tidal range was identified as the best alternative and ocean current the worst, among the projects evaluated.

Keywords: Multicriteria Decision Making Methods · Ocean energy · Renewable energy

1 Introduction

Globally, in the last decade, a continuous interest in cleaner, cheaper, and scaled up renewable energy alternatives for electricity generation has taken place, considering that an accelerated demographic growth has boosted fossil fuels consumption and affected air quality [1]. The key is to guarantee the quality of life of people and restrain the negative environmental impact of fossil fuels exploitation by promoting matured renewable energy technologies [2].

© ICST Institute for Computer Sciences, Social Informatics and Telecommunications Engineering 2021
Published by Springer Nature Switzerland AG 2021. All Rights Reserved
J. L. Afonso et al. (Eds.): SESC 2020, LNICST 375, pp. 63–83, 2021.
https://doi.org/10.1007/978-3-030-73585-2_5

To improve energy security programs, OECD countries have been exploring multiple alternatives for the sustainable exploitation of renewable energy sources, such as biomass, hydro, solar, wind, geothermal, ocean current, waves, thermal gradients, salinity gradients, seismic or vibrational, waste and green hydrogen [3]. Through technology and innovation, these kinds of energy sources have been promoted and exploited. In 2018, renewable energy technologies supplied about 26% of the global electricity production [4]. According to REN 21, renewable energy contributing to the final energy consumption supply increased 4% annually between 2013 and 2018, about 7,3 EJ. Among them, 48% of the sources exploited were solar and wind, by photovoltaic cells and wind turbine technologies, respectively [5].

Developing countries are highly dependent on fossil fuels. This dependence requires a thorough analysis to assess the feasibility of the implementation of renewable energy sources using decision making methods which allow to choose the best technological alternatives by identifying financial risks and emissions reduction. In Colombia, 68% of electricity consumption is supplied by hydroelectricity, 31% is supplied by natural gas and coal, and 1% is supplied by renewable energy sources such as solar and wind in the northern region. Considering current issues in Colombia, such as constant high rates of demographic growth, financial challenges, and its environmental commitments of COP 21 in Paris, it is necessary to find strategies for the development of scenarios to expand alternative energy sources. In order to accomplish said goals, the Mining and Energy Planning Unit (UPME) proposed a promotion program for renewable energy technologies with a goal of supplying 17% of the total energy consumption by 2030 [6, 7]. Exploiting renewable energy technologies such as solar photovoltaics, wind turbines, and new renewable energy technologies not included in Law 1715 of 2014, such as ocean energy, vibrational energy, and green hydrogen, could be a way to accomplish that goal.

This study shows a methodology that identifies the feasibility of renewable energy projects by considering technical, financial, and environmental indexes and selects the best alternative among a set of renewable energy technologies by Multicriteria Decision Making Methods (MCDM). Using this methodology, it is possible to compare power generation projects considering technical, financial, and environmental criteria, such as power capacity, technological readiness, investment costs, capacity factor, the reduction of GHG emissions, and Levelized Cost of Energy (LCOE). The methodology was applied to conduct a comparative assessment of ocean energy technologies with solar and wind projects in Colombia in order to establish parameters for developing energy transition strategies aiming toward cleaner energy sources. This kind of comparative theoretical assessment considering ocean energy technologies has yet to be developed in Colombia. Unlike similar traditional methodologies for conducting the assessment of renewable energy technologies for a region by MCDM, this proposed methodology was designed considering the simultaneous computation of three MCDM. In order to validate the results, the integration of technical, financial and environmental indexes was considered so as to adjust the numerical value of criteria in the decision matrix. A complementary algorithm was attached to the main code of the MCDM developed in Python to conduct a financial and performance analysis. This characteristic allows for the establishment of competitiveness thresholds based on the new values of criteria that are being computed.

2 Energy Sources

In the world, energy sources are classified as renewable energy sources and nonrenewable energy sources, according to their availability and the available technologies to exploit them. Renewable energy sources, also known as nonconventional energy sources, have unlimited potential, as these resources are the result of the transformation of solar radiation or the gravitational attraction among planets. In contrast, industrialized countries are highly dependent on nonrenewable resources, also known as conventional energy sources, such as coal, oil and natural gas. According to TRL, nonconventional energy sources such as solar, wind, biomass, and ocean energy, as opposed to conventional energy sources, generally are exploited by non-well-developed technologies at big scales and have low participation in the energy matrix [8]. Likewise, energy sources can be classified on primary and secondary sources. Primary energy sources are obtained directly from nature such as fossil fuels, solar, wind, geothermal and biomass and secondary energy sources, also known as energy vectors(such us hydrogen), are used for storage and transportation of energy [9].

The energy matrix of OCED countries, such as Iceland and Norway, is composed by a high percentage of renewable energy sources, due to the fact these countries have identified, in the last decade, the urgency for developing energy transition strategies aiming toward cleaner energy sources due to climate change issues. Nevertheless, the participation of coal, oil and natural gas continues to be extremely high in most countries. So, the necessity for better, wider, and more effective strategies to reduce the gross annual contaminant and GHG emissions have been identified.

In Colombia, Law 1715 of 2014 defines which energy sources are considered nonconventional. The objective of this law is the promotion of flows of capital on renewable energy projects, which guarantee financial development and environmental sustainability. This law establishes that in Colombia these kinds of resources must be sustainable and must not be widely commercialized and used for large-scale power generation. Nuclear energy, biomass, geothermal, small-scale hydro facilities, wind, solar, and ocean energy are considered nonconventional and are covered by incentives and grants [10]. Today, the Mining and Energy Planning Unit (UPME) of Colombia must identify and evaluate nonconventional energy sources that were not mentioned in the law to include them, if necessary. However, the inclusion of these energy sources necessitates financial and sustainable development, which allow for the guarantee of energy security and energy supplies in all regions. So, the condition that nonconventional energy technologies must be matured and commercialized widely abroad has been identified.

Among the energy sources mentioned by the Law 1715 of 2014 in Colombia, thermal gradients, salinity gradients, waves, tidal and ocean currents have not yet been considered through financed projects despite the existence of their energy potential. On the other hand, geothermal and nuclear energy are sources highly promoted by industrialized countries, but in Colombia the financial, technological, and optimal conditions for their scaled-up implementation have not been established. To classify the best non-conventional energy technologies available for Colombia, the Technology Readiness Levels (TRL), a classification scale of technologies designed by NASA, were considered. Though hydrogen is generally exploited by fuel cells used in power generation, buildings and light duty vehicle applications with TRL 5 to TRL 9, it was identified that the direct comparison with primary energy sources such as solar and wind is not an optimal standard [11, 12]. Unlike fuel cells, vibrational energy technologies were not considered in the assessment due to its TRL 4, an emerging readiness level [13, 14]. This classification is shown in Fig. 1.

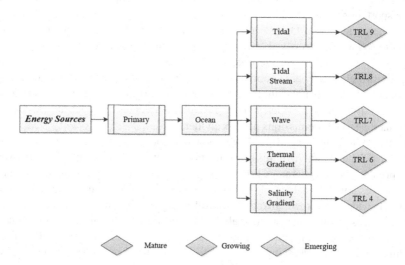

Fig. 1. Technology Readiness Levels (TRL) for ocean energy technologies.

In Colombia, geothermal energy is being evaluated by researchers, and the possibility to initialize its promotion, financing, and research is being considered. However, similar efforts have not been identified for ocean energy, despite it being an energy source with potential in the northern and western regions of Colombia, which could be exploited by technologies with high TRL levels, as presented in Fig. 1. For these reasons, ocean energy technologies considering waves, tidal, and ocean currents were chosen to conduct the assessment [15, 16].

3 Multicriteria Decision Making Methods (MCDM)

According to Ilbahar E, Cebi S y Kahraman C, the Multicriteria Decision Making Methods often used for the evaluation of renewable energy technologies include: *Analytical Hierarchy Process* (AHP), *ELimination Et Choix Traduisant la REalité* (ELECTRE) and *Technique for the Order of Preference by Similarity to Ideal Solution* (TOPSIS). Using these methods, it is possible to set alternatives and integrated strategies for electricity generation, identify optimal locations for infrastructure, and choose the best technological option.

Considering that MCDM can be classified into three groups, Elementary Methods, Unique Criteria Synthesis Methods, and Overcoming Methods, it is necessary to identify which group and which method contain the best characteristics for conducting the evaluation of renewable energy technologies in Colombia. Elementary methods, such as Dominance, *Weighted Sum Model* (WSM), *Weighted Product Model* (WPM), and Weighted Aggregates Sum Product Assessment– WASPAS – are characterized as low-complexity algorithms. Unique Criteria Synthesis methods, such as *Analytic Hierarchy Process* (AHP), *Technique for the Order of Preference by Similarity to Ideal Solution* (TOPSIS), and *VIšekriterijumsko KOmpromisno Rangiranje* (VIKOR) are characterized as medium-complexity algorithms by which it is possible to find optimal and non-optimal solutions. Overcoming methods, such us *Preference Ranking Organization METHod for Enrichment Evaluation* (PROMETHEE) and *ELimination Et Choix Traduisant la REalité* (ELECTRE), are characterized as being algorithms from which it is possible to establish dominance relations among alternatives [17–20].

Using a comparative analysis that considers characteristics such as Ranking, Attribute, and Criteria Weight Calculation, it is possible to select the most appropriate method for the evaluation.

- Ranking: capacity of the algorithm to determine the order of importance among the alternatives by an index which is computed.
- Attribute: capacity of the algorithm to assign the quality positive/negative, desirable/undesirable, or benefit/cost of the criteria. Positive attribute allows criteria to enhance the value of the alternative as much as its value increase. Power capacity and gross annual GHG emission reduction rate are considered positive criteria. On the other hand, Capital Cost Expenditure (CAPEX) and the Levelized Cost of Energy (LCOE) are considered negative criteria.
- Weight: by this criterion, it is possible to assign a dominance value among the selected criteria. Some methods include its computation within the algorithm. This value defines the relative importance among criteria.

In Table 1, the main characteristics of the MCDM considered as criteria for the selection are presented.

Table 1. Characteristics of MCDM algorithms.

MCDM		Ranking	Attribute	Weight
Elementary	Dominance	No	No	Yes
	WSM	Yes	No	Yes
	WPM	Yes	No	Yes
Unique criteria Synthesis	AHP	Yes	No	No
	TOPSIS	Yes	Yes	Yes
	VIKOR	Yes	Yes	Yes
Overcoming	ELECTRE	No	No	Yes
	PROMETHEE I	No	Yes	Yes
	PROMETHEE II	Yes	Yes	Yes

From Table 1, it is observed that TOPSIS, VIKOR and PROMETHEE II contain the capacity of making ranking for the alternatives. These require the definition of attribute for each weight, and the calculation of weights were conducted by an external algorithm, opening the possibility for objective weights. In this study, TOPSIS, VIKOR and PROMETHEE II are computed simultaneously by a code elaborated in Python in order to validate the output ranking. It was found that the assessment could be complemented by the computation of PROMETHEE I, from which an analysis of dominance for the alternatives was conducted.

3.1 Definition of Decision Matrix and Computation of Initial Weights

The calculation of weight w_j can be divided into subjective weight and objective weight. Subjective weight is mainly determined by an expert opinion based on experiences and subjective judgements. In contrast, objective weight is directly drawn from the real data of the alternatives in the decision matrix. An objective method, such as Shannon's entropy, reduces the impact of decision making and increases objectivity [21]. According to the entropy theory, the lower the entropy value, the more the information can be provided. A criterion can be assigned a greater weight if the difference among its values for each alternative is wider. A small numerical difference among values of a criterion for each alternative means a lower probability of obtaining information, higher entropy, and lower weight. The sum of weights must be equal to 1. The computation of Shannon's entropy weight is computed as follows:

a. Assuming m alternatives $(A_1, A_2, \ldots, A_m)$ and n criteria $(C_1, C_2, \ldots, C_n)$, the initial decision matrix is A, as presented in Eq. (1).

$$A = \begin{bmatrix} a_{11} \ a_{12} & & a_{1n} \\ & \cdots & \\ & \ddots & \\ & \cdots & \\ a_{n1} \ a_{n2} & & a_{mn} \end{bmatrix} = \left[a_{ij} \right]_{mxn} \tag{1}$$

where its elements a_{ij} denote i the alternative of j th criterion.

b. Normalize the decision matrix, where its elements $\hat{r}_{ij}$ are defined as presented in Eq. (2).

$$\hat{r}_{ij} = \frac{a_{ij}}{\sum_{i=1}^{m} a_{ij}}, i = 1, 2, \ldots, m \tag{2}$$

c. Compute entropy as presented in Eq. (3).

$$e_j = -\frac{1}{\ln(m)} \sum_{i=1}^{m} \hat{r}_{ij} \ln \hat{r}_{ij}, j = 1, 2, \ldots, n \tag{3}$$

d. Calculate the weight of each criterion as presented in Eq. (4).

$$w_j = \frac{1 - e_j}{\sum_{i=1}^{n} (1 - e_j)}, j = 1, 2, \ldots, n \tag{4}$$

4 Methodology

The methodology proposed, aligning with the Colombia energy market, should contain the following characteristics [7]:

a) A tool that supports the development of public policy.
b) The ability to conduct a comparative assessment of the renewable energy technologies considering technical, financial, and environmental criteria simultaneously.
c) The ability to assess multiple scenarios associated with variations of energy markets.

In this study, a methodology for the evaluation of renewable energy technologies in Colombia is proposed. This methodology has been validated by the Mining and Energy Planning Unit (UPME) of Colombia. Figure 2 presents the detailed and schematic diagram of the methodology. This proposal consists of three phases being conducted in the following sequence: a) Mapping the problem, b) Performance assessment and c) Multicriteria Decision Making Method (MCDM).

In Phase I, the evaluation case is defined, and the technical, financial, environmental, social and/or sociopolitical criteria are selected considering which criteria are most used for the assessment of renewable energy technologies. The criteria is selected by the AHP method which considers the judgement of experts from private and public sectors. The scenarios are established, and the assignation of attribute for criteria is conducted.

In Phase II, the financial and technical parameters for each project are defined, such as the electricity exported to grid, electricity export revenue, debt, and equity. Moreover, there are defined financial indexes, such as the Net Present Value (NPV), the Benefit-Cost Ratio (BCR), the Annual Life Cycle Savings (ALCS), the Simple Payback Period (SPP), the Equity Payback (EP), and the Levelized Cost of Energy (LCOE). Additionally, environmental indexes related to GHG such as the Gross Annual GHG emission reduction (GHG_{gr}) are assessed. Based on these indexes a feasibility assessment is conducted.

Note that these indicators are highly dependent on the sales price of energy, *"Annual Rate"*. In other words, there is a guarantee that the projects for this initial simulation are financially viable, given that the sales price of energy for the solar and wind project was calculated at 100 USD/MWh, and the ocean wave, tide, and currents project were calculated at 300 USD/MWh/ 200 USD/MWh, and 600 USD/MWh, respectively. The assumption is that the price at which each project can sell energy is within a reasonable range, which allows them to be competitive within the national energy market. However, despite the fact the levelized cost of energy (LCOE) might appear attractive when compared with estimated typical mean values for a specific technology, it may not be the case due to differences in the order of magnitude in the sales price of energy.

In phase III the comparative assessment of renewable energy technologies for the evaluation case and scenarios is conducted. The weights of criteria are computed by a method based on the calculation of Shannon Entropy. So, by the application of the Multicriteria Decision Making Methods (MCDM), the best and the worst alternatives from the output ranking are identified. Finally, the criteria which cause more variability on decision making by a sensitivity analysis based on the evaluation of scenarios are identified.

Note that criteria used in the MCMD to evaluate alternatives are as follows: Technological Readiness – TRL, Power Capacity – PC, Investment costs – IC, Levelized Cost of Energy – LCOE, and Emission Reduction – ER. Previously, in Phase II, financial indicators with which it is possible to execute an analysis of the financial performance of the projects are used. This allows for the identification of the specific sales price value of energy that guarantees the alternatives that are evaluated in the method.

On the other hand, even though some of the selected criteria to execute the evaluation are interdependent, like LCOE, the capacity factor, and CAPE Diverse studies that have selected these interdependent criteria did not assign importance to this dependence.

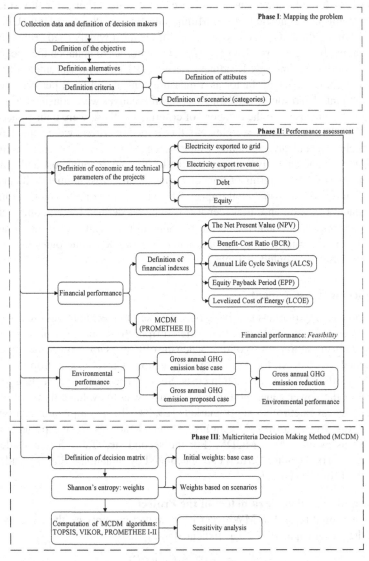

Fig. 2. Detailed steps and schematic diagram of the methodology.

4.1 Mapping the Problem

Definition of Criteria, Attributes, and Scenarios
Previously, the objective, the decision makers, and the alternative energy sources were identified Moreover, according to literature, the evaluation criteria for renewable energy can be divided into four main categories: financial, technical, environmental, and social dimensions. Criteria such as capital cost, O&M cost, Levelized Cost of Energy (LCOE), power capacity, efficiency, job creation/welfare improvement, social acceptability, emission reduction, and land use are commonly considered in renewable energy technologies studies. Among these, capital cost, Levelized Cost of Energy (LCOE), and the gross annual GHG emission reduction were chosen as the criteria for this studied in accordance with the suggestion of the Mining and Energy Planning Unit (UPME) of Colombia. Among these, GHG emission is one of the most widely used criteria in evaluating sustainability of renewables [22]. However, in order to enhance the spectrum of the assessment, the Technology Readiness Levels (TRL) as a measurement of technological readiness the capacity factor as a measurement of technical performance, and the energy potential and source availability of the alternatives were considered. The scenarios were chosen based on the category of criteria, and these were defined according to the numerical values of their weights. So, the following criteria were considered: the financial, technical, and environmental scenarios regarding power capacity; Technology Readiness Levels and capacity factor regarding technology; capital cost and Levelized Cost of Energy (LCOE) in terms of financial criteria; and gross annual GHG emission reduction in terms of environmental impact. The weights of criteria which define them were calculated based on an initial scenario which was defined by the computation of Shannon's entropy. Additionally, a scenario where all the weights own the same value was considered. So, criteria for the assessment of renewable technologies in Colombia based on MCDM is presented in Table 2.

4.2 Performance Assessment

The renewable energy markets involving power generation technologies are not usually balanced. In consequence, energy prices offered from those sources can increase over the total cost during short periods of time when the offer does not grow at the same speed as the demand. In contrast, for periods of surplus, losses can cause a drop in the price of the energy offered below production costs. So, it is necessary to execute a financial assessment of the power generation technology as a way to evaluate its feasibility and its competitiveness in current energy markets [23, 24]. To accomplish that, a two-step subroutine was proposed: a) calculation of financial indexes for the projects and b) comparative assessment by a MCDM. For the second step Table 5 is considered as the decision matrix, weight of criteria are calculated by Shannon's Entropy and data is computed by PROMETHEE II.

Economic and Technical Parameters of the Project
Electricity exported to grid in MWh/year is computed multiplying the capacity factor (CF) and the power capacity (CP) as indicated in Eq. (5).

$$E_{xg} = 0,0876(CF)(CP) \tag{5}$$

Table 2. List of criteria for the assessment based on MCDM.

Criteria		Scenario	Attribute	Unit
TRL	Tech. Readiness	Technical	+	–
PC	Power capacity		+	kW
CF	Capacity factor		+	%
IC	Investment cost	Financial	–	$/kWh
LCOE	LCOE		–	$/MWh
ER	Emission reduction	Environmental	+	tCO$_2$/year

*TRL: Technological Readiness Levels, PC: Power capacity, CF: Capacity factor, IC: Investment cost, LCOE: Levelized Cost of Energy and ER: Gross annal GHG emission reduction.

Electricity export revenue in USD/year is computed multiplying the electricity export rate and the electricity exported to grid as indicated in Eq. (6). However, it is escalated at the electricity export escalation rate (r_s).

$$E_{xr} = r_e E_{xg} \tag{6}$$

Debt is the fraction of the total investment required for the implementation of the project, and it is financed by a loan. The project debt leads to the calculation of the debt payments and the Net Present Value (NPV). It is calculated multiplying the total initial cost (IC) and the Debt Ratio (f_d). Debt is computed as indicated in Eq. (7)

$$D_b = f_d IC \tag{7}$$

On the other hand, equity is the fraction of the total investment required to finance the project that is funded directly by the facility owners. It is computed as indicated in Eq. (8).

$$E_q = (1 - f_d)IC \tag{8}$$

Financial Performance Assessment

The financial feasibility study of the projects considered is carried out by indexes such as the equity payback, the net present value, the yearly positive cash flow, the benefit-cost ratio, the annual life cycle savings, and the Levelized Cost of Energy [25].

The Net Present Value (NPV) of the project is the difference between the sum of cash inflow and outflow. It is computed by discounting all cash flow, as indicated in the Eq. (9).

$$NPV = \sum_{i=0}^{N} \frac{\hat{C}_n}{(1+r)^n} \tag{9}$$

where,

$\hat{C}_n$: Net cash flow during the period N.

r: Discount rate of the project.

N: Life of the project in years.

The Benefit-Cost Ratio (BCR) provides a measure of the financial desirability of the project. It is expressed as the ratio of the net benefits to the costs of the project. It is computed as indicated in Eq. (10).

$$BCR = \frac{NPV + (1 - f_d)C}{(1 - f_d)C} \tag{10}$$

where,
f_d: Debt ratio.
C: Total initial cost of the project.
The Annual Life Cycle Savings (ALCS) is the levelized nominal yearly savings, having the same life and net present value as the project. It is computed as indicated in Eq. (11).

$$ALCS = \frac{NPV}{\frac{1}{r}\left[1 - \frac{1}{(1+r)^N}\right]} \tag{11}$$

The Simple Payback Period (SPP) is the period required for the cash flow to be equal to the total investment. It is computed as indicated in Eq. (12).

$$SPP = \frac{C - IG}{(C_e + C_{cap} + C_{RE} + C_{GHG}) - (C_{O\&M} + C_{comb})} \tag{12}$$

where,
C: Total initial cost of the project.
IG: Incentives and grants.
C_e: Annual energy savings.
C_{cap}: Annual capacity savings.
C_{RE}: Annual renewable energy production credit income.
C_{GHG}: Greenhouse gases reduction income.
$C_{O\&M}$: Annual operation and maintenance cost.
C_{fuel}: Annual cost of fuel or electricity.
The Equity Payback Period (EPP) is the period that represents the length of time that it takes for the owner of a facility to recuperate their initial investment (equity) out of

the cash flow generated by the project. The equity payback considers project cash flow from its inception as well as the leverage (level of debt) of the project, which makes it a better time indicator of the project merits than the simple payback. It is computed as indicated in Eq. (13).

$$EPP = A + \frac{\left|\hat{C}_c\right|}{\hat{C}_A} \tag{13}$$

where,

A: The last period number with a negative cumulative cash flow.

$\hat{C}_c$: Cumulative net cash flow at the end of period A.

$\hat{C}_A$: Net cash flow during the period following period A.

The Levelized Cost of Energy (LCOE) represents the electricity export rate required in order to have a Net Present Value (NPV) equal to 0. It is computed as indicated in Eq. (14).

$$LCOE = \sum_{t=1}^{n} \frac{I_t + M_t + F_t}{(1+r)^t} \left[\frac{E_t}{(1+r)^t} \right]^{-1} \tag{14}$$

where:

I_t: Annual investment cost.

M_t: O&M annual cost.

F_t: Annual cost of fuel.

E_t: Annual electricity generation, MWh.

r: Discount rate of the project.

Environmental Performance Assessment

The environmental performance study of the projects considered is carried out by calculating? the gross annual GHG emission reduction based on the GHG emission of the base and proposed case. The GHG emission of the base case represents the amount of GHG emitted for the base case system. It is computed by multiplying the emission factor of the base case (EF_{bc}) and the electricity exported to grid, as indicated in Eq. (15).

$$GHG_{bc} = EF_{bc}E_{xg} \tag{15}$$

The GHG emission of the proposed case represents the amount of GHG emitted for the proposed case systems. It is computed multiplying the emission factor of the proposed case (EF_{pc}) and the electricity exported to grid, as indicated in Eq. (16).

$$GHG_{pc} = EF_{pc}E_{xg} \tag{16}$$

The Gross annual GHG emission reduction is based on emissions of both the base case and the proposed case systems on an annual basis. It is computed as indicated in Eq. (17).

$$GHG_{gr} = GHG_{bc} - GHG_{pc} \tag{17}$$

Likewise, the Gross annual GHG emission reduction is commonly established as a percentage reduction. It is computed as indicated in Eq. (18).

$$GHG_{gr(\%)} = \frac{\left|GHG_{pc} - GHG_{bc}\right|}{GHG_{bc}} * 100\% \tag{18}$$

The decision matrix of Colombia's renewable energy technologies is presented in Table 3.

Table 3. Decision matrix on Colombia's renewable energy technologies [26–30].

	TRL	PC [MW]	CF [%]	IC [$/kWh]	LCOE [$/MWh]	ER [tCO$_2$/year]
Solar	9	86	23	995	46	67,062
Wind	9	19	38	1800	71	26,151
Wave	7	2	30	4900	306	2736
Tidal	9	250	30	3412	196	342,043
Current	8	4	35	11,466	594	6384

The initial weights of the criteria w_j computed by Shannon's entropy algorithm were $w_{TRL} = 0.0002$, $w_{PC} = 0.3392$, $w_{IC} = 0.1344$, $w_{LCOE} = 0.1426$, $w_{ER} = 0.3758$ and $w_{CF} = 0.0060$. These weights are computed based on the numerical values of criteria from Table 3. It was found that the weights of capacity factor and TRL are much smaller than the others. This could be explained considering that a small numerical difference among values of a criterion for each alternative means a lower probability of obtaining information, higher entropy, and lower weight. So, power capacity, investment cost, LCOE, and GHG emission reduction criteria are much better for decision making in this case. Nevertheless, capacity factor and TRL are important for the overall analysis considering that for Colombia this assessment is one of the first steps for the development of a baseline of renewable energy technologies and sources, such as ocean energy which is a renewable energy source not yet considered in the energy matrix.

5 Results

Based on the methodology presented in Fig. 2, the method PROMETHEE II was applied to the data presented in Table 5 in order to conduct a relative comparison of the financial performance of the projects. Likewise, MCDM methods (TOPSIS, VIKOR and PROMETHEE II) were applied to the data presented in Table 3. A sensitivity analysis was conducted based on the weight value of criteria for each scenario, adjusting the sum of criteria of the same category, proportionally. Objective weights of criteria were calculated by Shannon's entropy. It was found that gross annual GHG emission reduction was the most important criterion for the development of ocean energy in Colombia, according to Fig. 3, due to a wider difference among the values of criteria which implies a higher probability for obtaining information, lower entropy, and a greater weight. Unlike gross annual GHG emission reduction, TRL and capacity factor were the less representative criteria during the assessment. In Table 4, Table 5 and Table 6, the economic and technical parameters, the financial performance, and the environmental impact of the alternatives are presented, respectively. The parameters and the results of the financial and the environmental performance assessment were used to grant the numerical values of criteria, such as the LCOE and the gross annual GHG emission reduction. A debt ratio of 70%, a debt interest of 7%, a debt term of 15 years, a discount rate of 9%, and an escalation rate of 2% were considered for the performance assessment of the projects.

Table 4. Economic and technical parameters of the alternatives.

Alternative	E_{xg} [MWh/year]	E_{xr} [\$/year]	D_b [\$]	E_q [\$]
Solar	176,043	17,604,315	49,024,500	21,010,500
Wind	65,026	6,502,636	19,464,900	8,342,100
Waves	5256	1,576,800	6,860,000	2,940,000
Tidal range	657,000	131,400	597,100,000	255,900,000
Ocean current	12,264	7358	9,553,600	4,094,400

Table 5. Financial performance.

Alternative	EPP [years]	NPV [\$]	BCR	ALCS [\$/year]	LCOE [\$/MWh]
Solar	1.8	114,872,774	6.5	12,583,907	46
Wind	2.8	27,356,694	4.3	2,996,829	71
Waves	7.6	2,181,380	1.7	238,962	306
Tidal range	6.9	230,636,658	1.9	25,265,433	196
Ocean current	6.9	12,322,432	1.9	1,349,879	594

Computing the proposed two-step subroutine, the financial indexes for the projects were calculated. The value of the financial indexes for each project is presented in Table 5, which is considered the decision matrix. To conduct the comparative assessment, weight of criteria, in this case, the Net Present Value (NPV), the Benefit-Cost Ratio (BCR), the Annual Life Cycle Savings (ALCS), the Levelized Cost of Energy (LCOE) were calculated by Shannon's Entropy and data was computed by PROMETHEE II. It was found that tidal range has the best financial performance and ocean current the worst. It was identified that the ALCS was the most important criterion for financial performance assessment. On the contrary, the GHG emission of the base and proposed cases were calculated considering an emission factor of 0,548 tCO_2/MWh. The base case was computed considering the power generation by a natural gas thermal power plant, an electricity generation efficiency of 38% and T&D losses of 5%. In Fig. 3 and Table 6 the gross annual GHG emission reduction of the alternatives is presented.

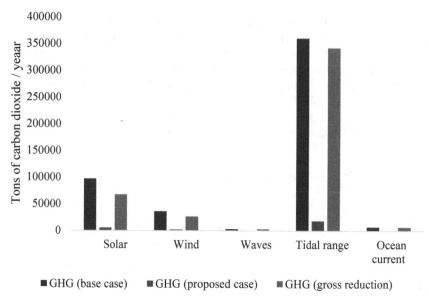

Fig. 3. Gross annual GHG emission reduction.

Table 6. Environmental performance.

Alternative	GHG_{bc}	GHG_{pc}	GHG_{gr}	$GHG_{gr(\%)}[\%]$
Solar	96,471	4823	67,062	95
Wind	35,634	1782	26,151	95
Waves	2880	144	2736	95
Tidal range	360,036	17,993	342,043	95
Ocean current	6721	337	6384	95

Based on the data of the decision matrix in Table 3, Shannon's Entropy was used to calculate the relative importance of each criterion. Since the criteria weight significantly cause changes on the rank, a sensitivity analysis was conducted to reveal the ranking alternatives changes due to variation of criteria weights. To accomplish that, the weight value of criteria for each scenario was adjusted to grant 80% of the sum of criteria of the same category, meaning, the remaining 20% for the other scenarios and criteria were reduced or increased proportionally from the base case to complete the sum equal to 1. In Table 7 the weights of the criteria under different scenarios is presented. In Fig. 4 and Table 8, the, in terms of different methods and scenarios, is presented.

Table 7. Criteria weights under different scenarios.

	TRL	PC	CF	IC	LCOE	ER
Base case	0.002	0.339	0.006	0.134	0.143	0.376
Equal	0.167	0.167	0.167	0.167	0.167	0.167
Technical	0.005	0.782	0.014	0.067	0.067	0.067
Financial	0.500	0.500	0.500	0.388	0.412	0.500
Environmental	0.040	0.040	0.040	0.040	0.040	0.800

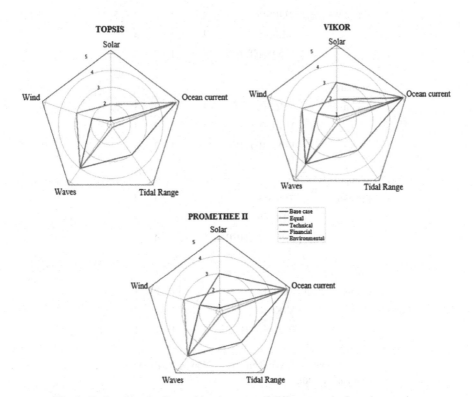

Fig. 4. Radar chart for the ranking in terms of different methods and scenarios.

Table 8. Ranking in terms of different methods and scenarios.

	Method	Rank				
Scenario 1		I	II	III	IV	V
Base case	TOPSIS	4	1	2	3	5
	VIKOR	4	1	2	3	5
	PROMETHEE II	4	1	2	3	5
Scenario 2						
Equal weight	TOPSIS	4	1	2	3	5
	VIKOR	4	1	2	3	5
	PROMETHEE II	4	1	2	3	5
Scenario 3						
Technical	TOPSIS	4	1	2	3	5
	VIKOR	4	1	2	3	5
	PROMETHEE II	4	1	2	3	5
Scenario 4						
Financial	TOPSIS	1	2	4	3	5
	VIKOR	1	2	4	3	5
	PROMETHEE II	1	2	4	3	5
Scenario 5						
Environmental	TOPSIS	4	1	2	3	5
	VIKOR	4	1	1	5	3
	PROMETHEE II	4	1	2	3	5

*1. Solar energy, 2. Wind energy, 3. Waves, 4. Tidal range, 5. Ocean current.

6 Conclusions

The assessment of renewable energy technologies can be conducted by Multicriteria Decision Making Methods considering technical, financial, and environmental criteria as a tool to reduce financial risks and guarantee the objective goals of renewable energy projects. The comparative assessment of ocean energy technologies with renewables such as solar and wind in Colombia was considered. A methodology based on three phases was proposed: I) Mapping the problem, II) Performance assessment, III) Multicriteria Decision Making Method.

During Phase I, there were five defined alternatives: solar, wind, waves, tidal range, and ocean current, and six defined criteria: Technological Readiness Levels, TRL; Power Capacity PC; Capacity Factor, CF; Investment Cost, IC; Levelized Cost of Energy, LCOE; and Gross Annual GHG Emission Reduction, ER. TRL, PC, CF and ER criteria were considered positive as opposed to IC and LCOE, which were considered negative. Likewise, there were five defined scenarios: base case, equal weight, technical, financial, and environmental. For the base case, Shannon's Entropy was used to calculate the relative importance of each criterion. The weights of criteria for the equal weight, technical, financial, and environmental scenarios were calculated adjusting the value of criteria to grant 80% of the sum of criteria of the same category.

In Phase II, financial and environmental assessment of the projects were conducted, and parameters such as electricity exported to grid, electricity export revenue, debt, equity and financial indexes such as Net Present Value, Benefit-Cost Ratio, Annual Life Cycle Savings), Levelized Cost of Energy (LCOE), and gross annual GHG emission reduction were calculated. The results of the financial and the environmental performance assessment were used to grant the numerical values of criteria such as the LCOE and the gross annual GHG emission reduction. It found that the gross annual GHG emission reduction percentage is the same for all the alternatives, but tidal range exhibits increased gross annual emission reduction due to its high-power capacity. Likewise, tidal range exhibits the best financial performance among the alternatives. Unlike tidal range, ocean energy exhibits the worst environmental and financial performance, as presented in Table 8.

Finally, in Phase III, data was computed by three MCDM (TOPSIS, VIKOR, and PROMETHEE II). The ranks obtained from the methods considered were similar, although not entirely equal. However, this issue points out the validity of the algorithms used. According to Table 8 and Fig. 4, tidal range is the best alternative and ocean current the worst for all the scenarios. This could be explained considering the difference in the power capacity, being higher for tidal range. Since higher power capacities eventually imply higher electricity exported to grid for similar capacity factors, and higher electricity exported to grid leads to higher gross annual GHG emission reduction for similar GHG emission factors and lower Levelized Cost of Energy for similar investment cost values. In this study, power capacity is the determining criteria in the assessment. Due to the fact tidal range exhibits the highest power capacity, it would be expected that this alternative was the best. Finally, it can be concluded that for similar power capacities, solar energy would be a better alternative than tidal range considering that solar's financial performance was better than tidal range, even though the power capacity of solar was approximately 65% lower, as shown in Table 3 and Table 8.

References

1. Stern, A.G.: A new sustainable hydrogen clean energy paradigm. Int. J. Hydrogen Energy **43**, 4244–4255 (2018). https://doi.org/10.1016/j.ijhydene.2017.12.180
2. Mitsos, A., Asprion, N., Floudas, C.A., Bortz, M., Baldea, M., Bonvin, D., et al.: Challenges in process optimization for new feedstocks and energy sources. Comput. Chem. Eng. **113**, 209–221 (2018). https://doi.org/10.1016/j.compchemeng.2018.03.013

3. OECD - Better policies for better lives. Data - Organisation for Economic Co-operation and Development n.d. https://www.oecd.org/fossil-fuels/data/. Accessed 6 April 2020
4. REN 21 - RENEWABLES NOW. Global Overview n.d. https://www.ren21.net/gsr-2019/chapters/chapter_01/chapter_01/#fig_4. Accessed 18 July 2019
5. REN 21 - RENEWABLES NOW. Renewables 2020 - Global Status Report n.d. https://www.ren21.net/gsr-2020/chapters/chapter_01/chapter_01/. Accessed 1 Sept 2020
6. Unidad de Planeación Minero Energética - UPME. Plan Energetico Nacional Colombia: Ideario Energético 2050 2015:184
7. Gómez-Navarro, T., Ribó-Pérez, D.: Assessing the obstacles to the participation of renewable energy sources in the electricity market of Colombia. Renew. Sustain. Energy Rev. **90**, 131–141 (2018). https://doi.org/10.1016/j.rser.2018.03.015
8. Foro Nuclear - Foro de la Industria Nuclear Española. Energía y fuentes de energía n.d. https://www.foronuclear.org/es/energia-nuclear/faqas-sobre-energia/capitulo-1/115488-icomo-se-clasifican-las-fuentes-de-energia. Accessed 5 Mar 2020
9. U.S. Energy Information Administration (EIA). Sources of energy 2019. https://www.eia.gov/energyexplained/what-is-energy/sources-of-energy.php. Accessed 4 May 2020
10. República de Colombia - Gobierno Nacional. LEY No 1715 del 13 de mayo de 2014. TAPPI J 2014;13. https://doi.org/https://doi.org/10.32964/tj13.5
11. International Energy Agency (IEA). The Future of Hydrogen – Analysis - IEA n.d. https://www.iea.org/reports/the-future-of-hydrogen. Accessed 4 Apr 2020
12. Berger, R.: Global consulting. fuel cells and hydrogen applications for regions and cities. Consolidated Technol. Introduction Dossiers **1**, 108–123 (2017)
13. Gholikhani, M., Roshani, H., Dessouky, S., Papagiannakis, A.T.: A critical review of roadway energy harvesting technologies. Appl. Energy **261**, 114388 (2020). https://doi.org/10.1016/j.apenergy.2019.114388
14. Yang, Z., Zhou, S., Zu, J., Inman, D.: High-performance piezoelectric energy harvesters and their applications. Joule **2**, 642–697 (2018). https://doi.org/10.1016/j.joule.2018.03.011
15. Wang, J., Liu, Y., Chen, L., Tang, J.: Using the technology readiness levels to support technology management in the special funds for marine renewable energy. Ocean 2016 - Shanghai (2014). https://doi.org/10.1109/OCEANSAP.2016.7485608
16. International Renewable Energy Agency - IRENA. Ocean energy: technology readiness, patents, deployment status and outlook. Int. Renew. Energy Agency IRENA 76 (2014). https://doi.org/10.1007/978-3-540-77932-2
17. Wang, J.J., Jing, Y.Y., Zhang, C.F., Zhao, J.H.: Review on multi-criteria decision analysis aid in sustainable energy decision-making. Renew. Sustain. Energy Rev. **13**, 2263–2278 (2009). https://doi.org/10.1016/j.rser.2009.06.021
18. Pradhan, S., Indraneel, S., Sharma, R., Kumar, D., Nathuram, R.: Optimization of machinability criteria during dry machining of Ti-2 with micro-groove cutting tool using WASPAS approach Materials Today : Proceedings Optimization of machinability criteria during dry machining of Ti-2 with micro-groove cutting tool using WASPAS approach. Mater Today Proc (2020). https://doi.org/10.1016/j.matpr.2020.02.972
19. Fontana Viñuanes, M.: Métodos de decisión multicriterio AHP y PROMETHEE aplicados a la elección de un dispositivo móvil (2015)
20. García, J.I.: Las Líneas Estratégicas del Sector Hídrico en México en Materia de Investigación. Una Jerarquización Empleando el Método PROMETHEE, Desarrollo Tecnológico y Formación de Recursos Humanos (2009)
21. Lee, H., Chang, C.: Comparative analysis of MCDM methods for ranking renewable energy sources in Taiwan. Renew. Sustain. Energy Rev. **92**, 883–896 (2018). https://doi.org/10.1016/j.rser.2018.05.007

22. Ilbahar, E., Cebi, S., Kahraman, C.: A state-of-the-art review on multi-attribute renewable energy decision making. Energy Strateg. Rev. **25**, 18–33 (2019). https://doi.org/10.1016/j.esr. 2019.04.014
23. International Renewable Energy Agency. Renewable Power Generation Costs 2018 (2018)
24. Himri, Y., Merzouk, M., Kasbadji Merzouk, N., Himri, S.: Potential and economic feasibility of wind energy in south West region of Algeria. Sustain. Energy Technol. Assessments **38**, 100643 (2020). https://doi.org/10.1016/j.seta.2020.100643
25. Lezama-Nicolás, R., Rodríguez-Salvador, M., Río-Belver, R., Bildosola, I.: A biblio-metric method for assessing technological maturity: the case of additive manufacturing. Scientometrics **117**(3), 1425–1452 (2018). https://doi.org/10.1007/s11192-018-2941-1
26. World Energy Council. World Energy Perspective - Cost of Energy Technologies 2013:48. ISBN 978 0 94612 130 4
27. Atlantis Resources. MeyGen Tidal Energy Project Phase 1a – Progress Update (2016)
28. Australian Renewable Energy Agency - ARENA. The Tidal Turbine Reef (TTR) Feasibility Study (2018)
29. IRENA International Renewable Energy Agency. Wave energy technology brief. IRENA Ocean Energy Technol. Br. **4**, 28 (2014)
30. Pöyry Management Consulting. Levelised Costs of Power From Tidal Lagoons 2014:38

A Regression Model to Assess the Social Acceptance of Demand Response Programs

Paula Ferreira$^{(\boxtimes)}$ ⬤, Ana Rocha⬤, and Madalena Araújo⬤

ALGORITMI Research Center, University of Minho, Guimarães, Portugal
paulaf@dps.uminho.pt

Abstract. Residential demand response has been playing an important role in the low carbon energy system transition. Although this is not a new concept, the popularity of Demand Response (DR) programs is growing, driven by the increasing opportunities that emerged with smart grid appliances as well as by their potential to support the integration of variable renewables generation. The end-user plays a key role in the successful deployment and dissemination of these DR programs. This study aims to assess social awareness and acceptance of DR programs, based on a survey for data collection and complemented with the regression models. The results suggest that the economic determinants, contribution to environmental protection as well as the extent of urbanization are important motivating drivers, to be explored in the future to encourage the residential consumers' participation in DR programs.

Keywords: Demand response · Social acceptance · Heterogeneous choice model (oglm) · Ordered logit regression (ologit) · Residential consumers

1 Introduction

Residential demand response has been playing an important role in the low carbon energy system transition [1]. Demand Response (DR), involves achieving changes in energy usage by end-users' customers', for instance, shifting demand from peak to off-peak demand periods. This change can be achieved through price signal, providing financial incentives for shifting the electricity usage for the demand periods when the electricity price is lower, based on the higher share of RES for electricity generation; direct control [2]; and automation of appliances [3]. DR is also referred to as a potential driver for mitigating the challenges of reducing the intermittency of renewable sources by reducing demand at times of low renewable supply and increasing the demand at times a surplus of renewable energy is available.

DR is not a new concept. However, it still has a limited role and electricity supply and demand are mainly balanced by ensuring that generation, reserves, and network capacity are sufficient to meet demand [4]. The expected large-scale electrification of the transportation and heating sectors should have a significant impact on the energy consumption and has been creating a growing interest on demand flexibility for market

J. L. Afonso et al. (Eds.): SESC 2020, LNICST 375, pp. 84–93, 2021.
https://doi.org/10.1007/978-3-030-73585-2_6

players and energy policies. In this context, the residential consumers' participation in DR programs could play an important role in the electricity system management.

Several publications have explored household responsiveness to demand-side management. For, instance, [5] used ordered logistic regression to estimate the tiered electricity pricing system (TEP) effectiveness and the results suggest this system helps to reduce the electricity expenditures in China; another relevant aspect noted was a significant and negative association between the TEP effectiveness and income as this effectiveness tends to be reduced for high income groups. The regression results revealed that sociodemographic characteristics play an important role in improving the tiered electricity pricing effectiveness. Also for China [6], a binary logistic regression was used to demonstrate if the survey respondents were willing to accept the peak and off-peak time pricing. The socioeconomic characteristics and the level of knowledge on the topic were found to be significant for the acceptance of tiered pricing with females and elderly consumers showing higher acceptance. Another study, applied to the United Kingdom (UK), [7] using ordered logit regression, examined the willingness of the respondents to switch from flat-rate electricity tariff to ToUs tariff. The authors concluded that this willingness was driven by differences in loss-aversion characteristics and ownership of demand flexible appliances rather than by socioeconomic and sociodemographic characteristics.

In this paper, we propose a methodology to assess the social awareness and acceptance of DR programs, based on a survey for data collection in Portugal. From the collected data, the proposed regression model was derived, aimed at determining the most critical drivers to encourage domestic consumers' participation in DR programs and their level of acceptance using an ordered scale "totally disagree" "tend to disagree" "tend to agree" and "totally agree".

This paper is set out as follows: Sect. 2 is dedicated to describing both the data and methodology used. Section 3 presents the results of the various determinants for electricity usage delaying. Finally, in Sect. 4, the conclusions and future remarks are presented.

2   Data Sources and Methodology

2.1   Sample Data Sources

This study uses an empirical research method in order to assess the social awareness and acceptance of DR programs. We attempt in particular to address (i) the motivational factors to delay the electricity use (ii) the perceived flexibility of the residential electricity users through the quantification of the acceptance of delay on the use of appliances and (iii) willingness to accept the automatic control of the heating and cooling system. The data for this study was obtained from a survey conducted by phone during May and June of 2018, in Portugal. The survey was administered to residents randomly selected from 278 total number of Portuguese municipalities, covering both rural and urban areas. The analysis only considers Continental Portugal (i.e., excluding Azores and Madeira Islands). In total, 385 valid responses were obtained, which ensured a 95% confidence degree with a 5% margin of error. Table 1 presents a description of the variables used in the study.

Table 1. Description of variables encoded into the Stata software

	Variables	Variables assignment
Sociodemographic characteristic	Gender	Female = 1, Male = 2
	Age	[18–24] = 1, [25–44] = 2, [45–64] = 3, above 65 years old = 4
	Education	Low = 1; Medium = 2; High = 3
	Professional activity	Unemployed = 1, Student = 2, Posted worker = 3, Self-employed worker = 4, Retired = 5, Domestic worker = 6
	Household size	Numeric
	Urban/Rural	Rural = 1, Urban = 2
Knowledge and dynamism on electricity consumption	ToUs tariffs	No familiar with ToUs tariff = 0, Familiar with ToUs tariff = 1
	Reading meter	No regular meter reading = 0, Meter reading = 1
	Switch electricity supplier	No switch of electricity supplier = 0, Switch electricity supplier = 1
	Smart meter	No ownership of a smart meter = 0, Ownership of smart meter = 1
Motivational factors for energy management	Environmental factors	"Doesn't know/doesn't answer" = 0, " Totally disagree" = 1, "Tend to disagree" = 2, "Tend to agree" = 3 "Totally agree" = 4
	Reduce energy imports	"Doesn't know/doesn't answer" = 0, " Totally disagree" = 1, "Tend to disagree" = 2, "Tend to agree" = 3 "Totally agree" = 4
	Reduction of electricity bill	"Doesn't know/doesn't answer" = 0, " Totally disagree" = 1, "Tend to disagree" = 2, "Tend to agree" = 3 "Totally agree" = 4
	Recommendation	"Doesn't know/doesn't answer" = 0, " Totally disagree" = 1, "Tend to disagree" = 2, "Tend to agree" = 3 "Totally agree" = 4

2.2   Methodology

Figure 1 presents the modelling structure used in the study following five different stages:

Stage 1
Bearing in mind the purpose of this study, a logistic regression was applied. According to the classification of dependent variables, ordinal response with a meaningful sequential order, Ordered Logit regression (ologit), or Ordered Probit regression (oprobit), which are based on the cumulative probabilities of the response variable are the most suitable regressions to be applied. According to Ref. [8], logit regression has two main advantages: (i) simplicity – the equation of logistic distribution function is simple, while on the other and, the equation of probit distribution function contains unquantified integral and (ii) – the interpretation of the coefficients is directly presented as logarithms of chances (probability), while in the probit regression the interpretation of the coefficients is not direct. The logit and probit models are very similar in terms of predictive accuracy. Logit regression was then decided to be employed (highlighted in orange in 1st stage of Fig. 1).

Stage 2
After introducing the traditional ordered logit model, the assumption of the Ordered Logit

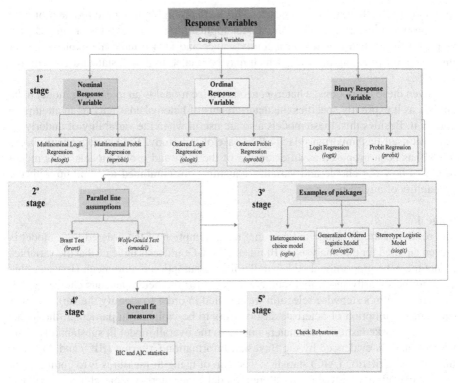

Fig. 1. Methodology for assessment the acceptance of DR programs (Color figure online)

regression was discussed in this second stage. According to the [9] this assumption is recurrently violated. The violation of the proportional odds/parallel lines assumption could lead to the formulation of an incorrect or mis-specified model. To validate the use of the Ordered Logit regression, we need to ensure the proportional odds/parallel lines assumption. For this purpose, the Brant and Wolfe-Gould tests were performed [10]. A Brant test provides both a global test to check whether any variables violates the proportional odd/parallel-lines assumption, as well as a test of the assumption for each variable considered [11]. Brant test suggested that the proportional odd/ parallel lines assumptions of the different dependent variables considered in the study was significantly violated (p-value) < .0.05). The only exception was for the dependent variable recommendation (p-value = 0.227), as p-value is higher than 0.05 the proportional odd/ parallel lines assumptions were not violated. The Wolf e-Gould test was used to confirm the results obtained by Brant test and led to similar conclusions.

Stage 3
Given the violation of the assumptions for the traditional ordered logit model, the possibility of using generalized ordered logit regression with a logistic cumulative distribution function was considered. However, according to Ref. [12] it is recommended to compute the predicted probabilities under gologit2 command in order to verify whether

this statistical technique is appropriate. This ends up highlighting the problem of negative probabilities, as a result of the model application to our data. Other studies such as [13] also reached negative probability values and [14] offer some explanation for this somehow puzzling outcome, which may be related to a high standard error on the responses.

Given these limitations, a heterogeneous choice model is an interesting model to be applied as it explicitly specifies the determinants of heteroskedasticity in an attempt to correct it. Besides this, these models also are useful when the variability of underlying attitudes itself has importance [15] as the case of this study. The heterogeneous choice model has been proposed as an extension of the logit and probit models. This model discloses how the choice and variance equations are combined to come up with the probability of any response.

Stage 4

The results of the 2nd stage indicate that the assumption of the ordered logit model is indeed violated for this analysis, as Brant and Wolf-Gould tests indicate that the variables do not meet the proportional odds/parallel lines assumption requirement. However, this does not necessarily mean that data are suitable for the heterogeneous choice model. Therefore, oglm's stepwise selection was applied in order to identify the variables that cause the assumption of heteroskedastic errors to be violated. In particular, the inclusion of heteroskedasticity parameters improves the overall model fit substantially. This improvement is evidenced by the Bayesian Information Criterion (BIC) and Akaike's Information Criterion (AIC) statistics. The aim of using fit measures is to compare the relative plausibility between two different models: the heterogeneous choice model and stereotype logistic regression (slogit) in order to find the best model (4th stage of Fig. 1). BIC measure evaluates the overall fit of the models. AIC measure is used to compare the models across the different samples [16]. These measures are defined as:

$$BIC = -2 * ln(LL) + 2 * k \qquad (1)$$

$$AIC = -2 * l(LL) + ln(N) * k \qquad (2)$$

Where LL is the log-likelihood, k is the number of parameters estimated and N is the number of observations.

The results pointed to the heterogeneous choice model as the most suitable one as highlighted in orange in 3rd stage of Fig. 1.

Stage 5

The results of the 5th stage use a common practice to the estimated model robustness by analysing if coefficients change the effect (positive or negative) when the regression model is modified. The comparison derived from the empirical analysis of the estimation of the three models: heterogeneous choice model (ogml) stereotype logistic regression (slogit) e generalized ordered logit regression (gologit2). As no significant difference was found on the estimated coefficients, the chosen model (heterogeneous choice model) could be considered robust and the results can be interpreted as a true casual effect between explained variable and explanatory variables.

3 Results

In this section, we present the models obtained from empirical analysis, which allow obtaining the response (dependent variable) predicted by the respondent's answers (independents variables). Based on the above mentioned, we conducted a heterogeneous choice model and ordered logit regression (which are both applied for ordinal dependent variables) using the commands oglm and ologit in the statistical software Stata 15. Accordingly, Table 2 shows the results of the estimated coefficients. The significance of the coefficients of variance equations may be relevant enabling to measure the variability attitudes towards end-user's participation in DR programs.

Table 2 discloses the results for the motivational factors for participating in energy management programs.

Regarding the environmental determinants, a positive coefficient in the variance equation suggests that respondents living in large households tend to present less disperse or variable attitudes towards participating in DR programs. On the other hand, in regarding the reduction of electricity bill determinant, a negative coefficient in the variance equation reveals that older people tend to present less disperse or variable responses when compared to young people. Moreover, the variability in attitude towards environmental benefit declined across the value of electricity bill meaning that respondents paying larger bills will tend to present less variable responses. Additionally, respondents who own a smart meter tend to present higher variability on the value assigned to environmental benefit; this could be explained by the fact that respondents who own a smart meter are more focused on financial incentives than to benefit the environment. The results presented by the equation of choice are interpreted the same way as a traditional logistic regression. Therefore, the results suggest that a large household and the knowledge of the possibility to shift from flat tariff to ToUs tariff influence positively the respondents do defer their electricity consumption motivated by environmental determinants.

The women and people who live in rural areas are more likely to accept to shift their electricity consumption encouraged by the potential financial gains. The negative coefficients in the variance equation reveal that the variability of the responses for older people and large household is lower than for younger respondents living in small households. A positive coefficient in variance equation suggests that the group of people who have knowledge on ToUs tariff and who regularly communicate with the electricity supplier tend to present more disperse responses in what concerns their attitudes towards deferring the electricity usage motivated by the financial issues. A large household could have a positive effect do defer the electricity consumption motivated by the contribution to reducing dependence on imported energy or by following acquaintance recommendation. The female gender and again the familiarity on ToUs are significant factors for the choice equation showing that these groups tend to be more sensitive to the energy dependence argument to participate in DR programs.

4 Discussion

The active role of women in electricity usage has been an object of analysis in other studies. However, this study found that women tend to be more active than men in the

Table 2. Aggregation analysis for motivational factors for energy management

Regression models		Heterogeneous choice model (oglm)			Ordered logit regression (ologit)
Motivational factors for energy management		Environmental factors	Reduction of electricity bill	Reduce energy imports	Recommendation
Equation CHOICE					
Sociodemographic determinants	Male	− 0.273	− 0.297*	− 0.537*	− 0.041
	Age	0.056	0.034	0.257	0.050
	Professional activity	0.072	0.072	0.083	0.093
	Education level	0.046	0.020	0.287	0.059
	Household size	0.360***	0.022	0.416***	0.145*
	Rural area	0.200	0.275*	0.296	0.138
	ToUs	0.483*	0.261	0.627**	0.434
Knowledge and dynamism of the respondents	Reading meter	− 0.149	− 0.048	0.357	− 0.020
	Switch electricity supplier	0.033	0.072	− 0.011	0.045
	Smart meter	0.740	0.346	0.886	0.331
Electricity bill value	Electricity bill value	− 0.135	0.054	0.218	− 0.011
Equation VARIANCE					
Sociodemographic determinants	Male	–	–	0.331***	–
	Age		− 0.330***		–
	Professional activity	–	–		–
	Education level	–	–		–
	Household size	0.114*	− 0.162**		–
	Rural area	–	–	0.411***	–
	ToUs	–	0.594***		–
Knowledge and dynamism of the respondents	Reading meter	–	0.348*		–
	Switch electricity supplier	–	–		–
	Smart meter	0.626**	–	0.506**	–
Electricity bill value	Electricity bill value	− .184**	–		–
Cut- points	/cut1	− 0.597	− 0.135		− 1.085
	/cut2	− 0.108	− 0.086		0.110
	/cut3	0.599	0.208		1.132
	/cut4	1.817***	0.688		2.541
	N	385	385		385
	Pseud R2	0.0327	0.0658		0.0082
	LR test (Chi2)	35.01***	60.48***		9.58
	Log-likelihood	−518.28626	−429.49223		−580.53897
	AIC		896.9845		1191.078
	BIC		972.0961		1250.377

Note: AIC – Akaike Information Criterion. BIC – Bayesian Information Criterion. The LR test tests the null hypothesis, which states that there was no difference between the model without independent variables and the model with dependent variables. ***, **, and * denote statistical significance level at 1%, 5% and 10%, respectively.

participation of DR programs, as shown in Table 2. The pertinence of these findings is particularly relevant in the Portuguese context, given that activities such as cooking and laundry represent a large share of energy consumption at the household level. These tasks are mainly performed by women, which should be taken into consideration in energy planning. Additionally, such results could also be used to foster shared responsibility at the household level, possibly contributing to balance the energy-related household chores and decision making.

Our results suggest that the household size could be a crucial driver to foster the DR programs, with larger households showing higher interest in participating in these programs as shown in Table 2. An additional member also could encourage positively to reduce electricity consumption. Namely, young children as a result of school education on energy and environmental protection principles could encourage parents and relatives towards a lower carbon lifestyle and more sustainable household patterns. This finding was also suggested by [17].

Another relevant outcome of this paper is the relevance of the monetary and environmental determinants to increase demand flexibility. Rather consensual responses were obtained, when questioned about the possibility to reduce the electricity bill, suggesting that the financial incentives are a crucial determinant for increasing the end-user flexibility demand motivated. This finding is also suggested by the statistical results that state that a large share of the respondent's answer is "totally agree" to defer the electricity usage motivated by the possibility to reduce the electricity bill.

5 Conclusions and Policy Implications

This current study is focused on the analysis of the social awareness and acceptance of DR programs, based on a survey for data collection complemented with statistical models into the residential sector. In this regard, two different regression models were estimated, the ordered logit regression and heterogeneous choice model, separately. The heterogeneous choice model was performed when the assumption of parallel lines was violated. This can help to avoid errors concerning the statistical significance of the explanatory variables.

From the resulting models, the role of women on electricity demand flexibility could be inferred. This finding is lengthily discussed to promote the participation of women in the decision making of the energy sector. Women are strongly associated with household chores, such as laundry, and the use of domestic appliances such as washing machines and dryers, which have a high potential to increase demand flexibility. The analysis also found that the level of higher education could increase the success of the DR programs, this group is related to the high use of new technologies, which is the first step towards a broad implementation of smart appliances that may support DR programs. Also, DR program implementation seems to be easily accepted by people living in urban areas which can create interesting synergies with the emergence of smart and sustainable cities. The analysis found that the flexibility is greatly linked to cost determinants. It is particularly important that the potential cost saving can somehow compensate for the inconvenience which may arise from the increase in flexibility with impacts on daily routines as well as in the comfort of the household. It is also noteworthy that environmental concerns play an important role in the willingness to participate in DR programs deployment and dissemination.

This study highlights could provide crucial information for energy policy and energy companies in order to define suitable strategies of development for further improvement on the power grid as well as encouraging the end-users to be more flexible. Moreover, the importance assigned to environmental and cost concerns should not be overlooked in the design of programs to increase the level of awareness on demand flexibility. This study should be seen as a first approach to design models that may explain the acceptance and willingness to participate in DR programs, but the complexity of the topic and related questions call attention to the need to proceed with further research on the topic. In particular, it would be important to extend the number of participants to allow for the use of different statistical techniques such as factorial or cluster analysis that could significantly contribute to the debate.

Acknowledgement. This work is financed by the ERDF – European Regional Development Fund through the Operational Programme for Competitiveness and Internationalisation COMPETE 2020 Programme, and by National Funds through the Portuguese funding agency, FCT Fundação para a Ciência e a Tecnologia, within project SAICTPAC/0004/2015-POCI/01/0145/FEDER/016434. As well as by the ALGORITMI research Centre within the R&D Units Project Scope: UIDB/00319/2020.

References

1. Parrish, B., Gross, R., Heptonstall, P.: On demand: Can demand response live up to expectations in managing electricity systems? Energy Res. Soc. Sci. **51**(January), 107–118 (2019)
2. Weck, M.H.J., van Hooff, J., van Sark, W.G.J.H.M.: Review of barriers to the introduction of residential demand response: a case study in the Netherlands. Int. J. Energy Res. **41**(6), 790–816 (2017)
3. Lopes, M.A.R., Henggeler Antunes, C., Janda, K.B., Peixoto, P., Martins, N.: The potential of energy behaviours in a smart(er) grid: policy implications from a Portuguese exploratory study. Energy Pol. **90**, 233–245 (2016)
4. Strbac, G.: Demand side management: Benefits and challenges. Energy Pol. **36**(12), 4419–4426 (2008)
5. Zhang, S., Lin, B.: Impact of tiered pricing system on China's urban residential electricity consumption: survey evidences from 14 cities in Guangxi Province. J. Clean. Prod. **170**, 1404–1412 (2018)
6. Yang, Y., Wang, M., Liu, Y., Zhang, L.: Peak-off-peak load shifting: are public willing to accept the peak and off-peak time of use electricity price? J. Clean. Prod. (2018)
7. Nicolson, M., Huebner, G., Shipworth, D.: Are consumers willing to switch to smart time of use electricity tariffs? The importance of loss-aversion and electric vehicle ownership. Energy Res. Soc. Sci. **23**, 82–96 (2017)
8. Kliestik, T., Ko, K., Mišanková, M.: Logit and Probit Model used For Prediction of Financial Health of Company, vol. 23, no. October 2014, pp. 850–855 (2015)
9. Williams, R.: Understanding and interpreting generalized ordered logit models. J. Math. Sociol. **40**(1), 7–20 (2016)
10. Long, J.S., Freese, J.: Regression Models for Categorical Dependent Variables Using Stata (2001)
11. Williams, R.: Gologit2: a program for Generalised Logistic Regression manual, p. 18 (2005)

12. Williams, R., Dame, N.: Gologit2 Documentation. Stat. J. **6**(1), 58–82 (2007)
13. Cheek, P.J., McCullagh, P., Nelder, J.A.: Generalized linear models. Appl. Stat. **39**(3), 385 (1990). 2nd Edn.
14. Keele, L., Park, D.K.: Difficult choices: an evaluation of heterogeneous choice models (2006)
15. Williams, R.: Fitting heterogeneous choice models with oglm. Stat. J. **10**(4), 540–567 (2010)
16. Williams, R., Dame, N.: Scalar measures of fit : pseudo R 2 and information measures (AIC & BIC) first we present the results for an OLS regression and a similar logistic regression . incbinary is a dichotomized version of income where the higher half of the cases are coded 1, p. 16 (2015)
17. Aguirre-Bielschowsky, I., Lawson, R., Stephenson, J., Todd, S.: Kids and Kilowatts: socialisation, energy efficiency, and electricity consumption in New Zealand. Energy Res. Soc. Sci. **44**(September 2017), 178–186 (2018)

Power Electronics; Power Quality

Efficiency Comparison of Different DC-DC Converter Architectures for a Power Supply of a LiDAR System

Ruben E. Figueiredo[1]($\boxtimes$), Vitor Monteiro[1], Jose A. Afonso[2], J. G. Pinto[1], Jose A. Salgado[1], Luiz A. Lisboa Cardoso[1], Miguel Nogueira[3], Aderito Abreu[3], and Joao L. Afonso[1]

[1] ALGORITMI Research Centre, University of Minho, Guimarães, Portugal
ruben.figueiredo@algoritmi.uminho.pt
[2] CMEMS-UMinho Center, University of Minho, Guimarães, Portugal
[3] Bosch Car Multimedia Portugal, S.A. Rua do Barrio de Cima, n°1, 4705-820 Braga, Portugal

Abstract. LiDAR (Light Detection And Ranging) is a technology used to measure distances to objects. Internally, a LiDAR system is constituted by several components, including a power supply, which is responsible to provide the distinct voltage levels necessary for all the components. In this context, this paper presents an efficiency comparison of three different DC-DC converter architectures for a LiDAR system, each one composed of three DC-DC converters: in parallel; in cascade; and hybrid (mix of parallel and cascade). The topology of the adopted integrated DC-DC converters is the synchronous buck Switched-Mode Power Supply (SMPS), which is a modified version of the basic buck SMPS topology. Three distinct SMPSs were considered: LM5146-Q1, LM5116, and TPS548A20RVER. These SMPSs were selected according to the requirements of voltage levels, namely, 12 V, 5 V, and 3.3 V. Along the paper, the principle of operation of the SMPSs is presented, as well as the evaluation results obtained for different operating powers, allowing to establish a comprehensive efficiency comparison.

Keywords: LiDAR · DC-DC converters · Efficiency comparison · Synchronous buck · Switched mode power supply

1 Introduction

Autonomous vehicles are identified as the main booster in terms of future intelligent transport in smart and sustainable cities [1]. Within this context, LiDAR (Light Detection and Ranging) is one of the key sensing technologies required to enable partial or fully autonomous driving [2]. Specifically, LiDAR is an active remote sensing method that works on the same principle of sonar, but using laser pulses to build a 3D model of the environment around. Internally, a LiDAR system is constituted by several parts such as a laser source capable of transmitting pulsed or continuous light, a low noise high-speed

J. L. Afonso et al. (Eds.): SESC 2020, LNICST 375, pp. 97–110, 2021.
https://doi.org/10.1007/978-3-030-73585-2_7

receiver capable of detecting and processing the reflected light beam and a low power controller unit, where the power supply is a common system to all the parts, which must guarantee the proper voltage and power levels. Concretely, the power supply is constituted by DC-DC power converters, each one with specific output voltage levels.

DC-DC power converters, in the perspective of an integrated solution, have increasingly been identified as extremely important, and arisen as an interesting and valid solution to the growing need for various voltage levels. The application of solutions based on integrated DC-DC converters presents a set of significant advantages over the solutions implemented using discrete DC-DC converters (e.g., less volume, weight and cost, and more robustness), mainly for automotive applications [3], such as LiDAR systems. It is also important to highlight that integrated DC-DC converters do not require large resources in terms of additional passive components, also allowing to increase the reliability and significantly simplifying their assembly and testing, when embedded in the target application.

In [4], it is proposed an innovative DC-DC power converter based on an integrated capacitive step-up structure, specifically dedicated to a LiDAR system in automotive applications. This converter operates from an input voltage of 12 V, allowing to supply the load with 70 V, with a maximum operating power of 320 mW. In [5], it is proposed the design and development of a power supply based on a DC-DC power converter for a high-power diode laser, for application on a LiDAR, which is part of an unmanned aerial vehicle. A new perspective of digital control, based on a variable-frequency with a multi-megahertz characteristic, is proposed in [6] for the application on a boost converter used on a LiDAR, presenting as main contribution the possibility to improve the efficiency, even when operating with sudden power variations.

Regarding general DC-DC power converters, several reviews can be found in the literature [7, 8], including DC-DC converters for very specific applications [9, 10]. However, for the most part, the presented solutions use many discrete components in the implementation of the DC-DC converters, not being the most beneficial for applications as LiDAR systems. In addition, in LiDAR systems, different architectures can be optionally used by associating distinct DC-DC power converters, when there is a need to supply different output voltages. For example, it is possible to identify in the literature architectures with parallel DC-DC converters (i.e., single-input multiple-output, SIMO), cascade DC-DC converters (i.e., the output of a converter is the input of the next one) and DC-DC converters with a hybrid association (i.e., combining parallel and cascade structures) [11, 12].

In this context, this paper presents a comparison between three different architectures for the power supply of a LiDAR, by reorganizing the same DC-DC converters in the three different architectures: (i) an architecture with all of them in parallel; (ii) an architecture with the three DC-DC converters linked in cascade; and (iii) a hybrid architecture, with two DC-DC converters connected in parallel to the output of the third one. The topology adopted in the scope of this paper is the synchronous buck Switched-Mode Power Supply (SMPS), which is a modified version of the conventional buck converter SMPS, in which the diode is replaced by a second power switch (MOSFET), minimizing the diode conduction losses and thus improving the conversion efficiency [16]. The LiDAR systems requires a total power of 50 W for the three different voltages values according to its function and divided as: 17 W at 3.3 V, 5 W at 5 V, and 28 W at 12 V. In terms of the value of the input voltage for the converters, the maximum voltage to be considered is 65 V, which is a requisite of the LiDAR system. For this purpose, the following models of DC-DC converters were considered: (i) the switching controller LM5146-Q1 [13]; (ii) the switching controller LM5116 [14]; (iii) the switching regulator TPS548A20RVER [15].

As the main contribution of this paper, it can be highlighted a comparative analysis, in terms of efficiency, among the three power supply architectures, all respecting the voltage limits of each DC-DC converter. The efficiency was determined for the highest voltage value possible at the input of each converter and with the maximum power required, usually, the most critical condition for real systems.

The rest of the paper is organized as follows. Section 2 presents the synchronous buck converter topology used by the selected converter chips. Section 3 describes the architectures proposed in this study, and Sect. 4 presents the main results obtained with each architecture, as well as the energy efficiency comparison for the three cases. Finally, Sect. 5 presents the main conclusions of the paper.

2 Synchronous Buck SMPS Topology

The synchronous buck SMPS topology is represented in Fig. 1, which allows to obtain a regulated output voltage lower than the input voltage. As shown, it is mainly constituted by two power MOSFETs, an input capacitor C1, an output inductor L1, and an output capacitor C2.

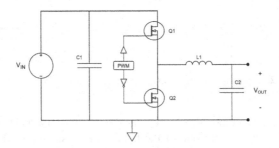

Fig. 1. Simplified schematic of the synchronous buck converter.

The average voltage at the output of the synchronous buck converter is directly influenced by the duty-cycle applied to the MOSFET gate terminals. Besides, some other important parameters, such as the current ripple, the voltage ripple and the minimum and maximum current through the inductor are dependent on the frequency and duty-cycle [17]. The synchronous buck converter has two operating modes. In mode 1, Q1 is 'on', while Q2 is 'off', therefore, the current in the inductor increases. In mode 2, Q2 is 'on', while Q1 is 'off', providing a path for the energy stored in the inductor to flow through the load. The two MOSFETs (Q1 and Q2) are alternately switched according to the PWM signal generated by a controller IC [18].

Comparatively to the non-synchronous buck SMPS topology, the synchronous buck topology has generally less power dissipation, resulting in a higher efficiency [17], since the voltage-drop across the low-side MOSFET (Q2) can be lower than the voltage-drop across the diode, in the case of the non-synchronous topology.

3 Architectures in Comparison

Automotive applications, such as the considered LiDAR system, require high efficient DC-DC conversions. In the particular application scenario of this study, the available input voltage ranges from 15 V to 65 V, higher than the required output voltages, 3.3 V, 5 V, and 12 V. Because of that, SMPSs (topology described at Sect. 2) were used to step down the input voltage to the intended values. Figure 2 presents the two standard architectures, parallel and cascade, while Fig. 3 presents the four possible hybrid architectures that could, in principle, also be adopted for the power supply satisfying this application.

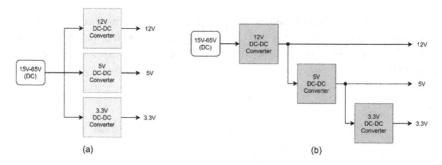

Fig. 2. Standard architectures: (a) Parallel (SIMO); (b) Cascade.

To demonstrate the methodology used in the efficiency comparison, and for the sake of simplicity, in this work only one of the hybrid variants, Fig. 3 (a), is used in the comparison with the parallel and cascade architectures, shown in Fig. 2. It was selected this architecture because it seems that it guarantees better efficiency since the voltage in the first stage is the nearest of the output voltage.

The input voltages, output voltages, output current, as well as the SMPSs used in each architecture, are presented in Table 1. LM5146-Q1 and LM5116 are switching controllers with a wide input voltage range (5.5 V to 100 V and 6 V to 100 V, respectively) and a wide adjustable output voltage (0.8 V to 60 V and 1.215 V to 80 V, respectively). TPS548A20RVER is a switching regulator with input range from 1.5 V to 20 V, output voltage from 0.6 V to 5.5 V and a maximum output current of 15 A.

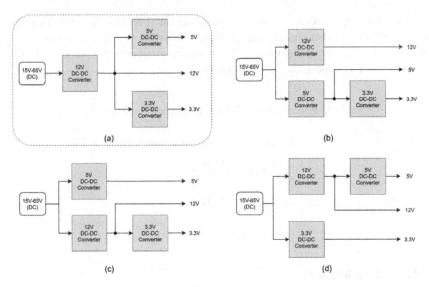

Fig. 3. Hybrid variant architectures.

Table 1. Specifications of the power supply characteristics of each architecture.

	12 V DC-DC Converter	5 V DC-DC Converter	3.3 V DC-DC Converter	Architecture
Input voltage	15 V–65 V	15 V–65 V	15 V–65 V	Parallel
Output voltage	12 V	5 V	3.3 V	
Converter Chip	LM5146-Q1	LM5146-Q1	LM5116	
Output current	2.33 A	1 A	5.15 A	
Input voltage	15 V–65 V	12 V	5 V	Cascade
Output voltage	12 V	5 V	3.3 V	
Converter Chip	LM5146-Q1	LM5146-Q1	TPS548A20RVER	
Output Current	4.24 A	4.47 A	5.15 A	
Input voltage	15 V–65 V	12 V	12 V	Hybrid
Output voltage	12 V	5 V	3.3 V	
Converter Chip	LM5146-Q1	TPS548A20RVER	LM5146-Q1	
Output current	4.22 A	1 A	5.15 A	

4   Evaluation Results

To evaluate and compare the performance of each architecture, the DC-DC convert-ers were simulated using the WEBENCH® Power Designer software [19], in order to determine its individual efficiency at an ambient temperature of 25°C. The efficiency was obtained for the maximum operating power required at each output voltage level, as shown in Table 2, also considering the highest possible voltage at the input, corresponding to the most critical situation. The switching frequency of each converter was selected in order to achieve the best performance.

Table 2. Power required by the LiDAR at each voltage level.

Voltage level	Operating power
3.3 V	17 W
5 V	5 W
12 V	28 W

4.1 Parallel Architecture

In this architecture the DC-DC converters (12 V, 5 V and 3.3 V) share the same input voltage, ranging from 15 V to 65 V, as illustrated in Fig. 2 (a).

According to the graph displayed in Fig. 4, the obtained efficiency of the 3.3 V DC-DC converter for an output current of 5.15 A and an input voltage of 65 V (worst case) is approximately 84.5%. For the same output current, but with lower input voltages, the efficiency is higher, namely, 88% for an input voltage of 40 V, and, approximately, 93% for an input voltage of 15 V.

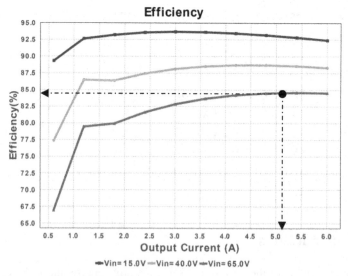

Fig. 4. Obtained efficiency regarding the output voltage level of 3.3 V (Parallel Architecture).

According to the graph displayed in Fig. 5, the obtained efficiency of the 5 V DC-DC converter for an output current of 1 A and an input voltage of 65 V (worst case) is approximately 75%. As in the previous converter, for the same output current, with lower input voltages, the efficiency is higher, namely, 82.5% for an input voltage of 40 V and 92% for an input voltage of 15 V.

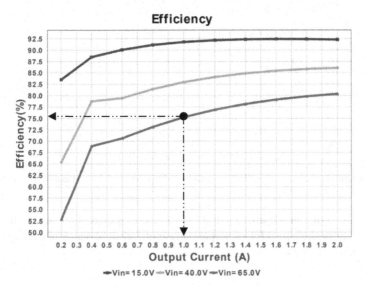

Fig. 5. Obtained efficiency regarding the output voltage level of 5 V (Parallel Architecture).

According to the graph displayed at Fig. 6, the expected efficiency of the 12 V DC-DC converter for an output current of 2.33 A and an input voltage of 65 V (worst case) is approximately 95.2%. Once again, for the same output current, with lower input voltages the efficiency is higher, namely, 96.2% for an input voltage of 40 V and approximately 98.2% for an input voltage of 15 V.

Therefore, the overall efficiency for the parallel architecture considering the most critical situation is:

$$\eta = \frac{50 \text{ W}}{\frac{5 \text{ W}}{0.75} + \frac{17 \text{ W}}{0.845} + \frac{28 \text{ W}}{0.952}} = 0.8897 = 88.97\% \tag{1}$$

where 50 W is the total output power and 5 W/0.75 is the 3.3 V DC-DC converter input power, 17 W/0.845 is the 5 V DC-DC converter input power and 28 W/0.952 is the 12 V DC-DC converter input power.

4.2 Cascade Architecture

In this architecture the three converters are connected in cascade; therefore, the 12 V DC-DC converter has the input range of 15 V to 65 V, the 5V DC-DC converter has the input of 12 V and the 3.3 V DC-DC converter has the input of 5 V, as illustrated in Fig. 2 (b).

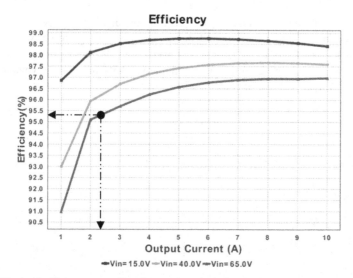

Fig. 6. Obtained efficiency regarding the output voltage level of 12 V (Parallel Architecture).

According to the graph displayed at Fig. 7, the expected efficiency of the 3.3 V DC-DC converter ($Eff_{C3.3V}$) for an output current of 5.15 A and an input voltage of 5 V is approximately 97.95%.

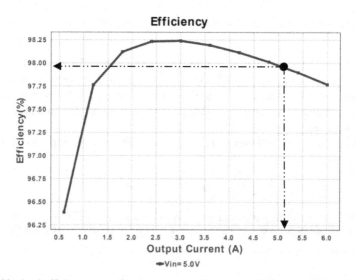

Fig. 7. Obtained efficiency regarding the output voltage level of 3.3 V (Hybrid Architecture).

The output current required at the 5 V DC-DC converter for the 5 W requested by the LiDAR system and to power the 3.3 V DC-DC converter is:

$$Iout_{5Vconverter} = \frac{5\ \text{W} + \frac{17\ \text{W}}{Eff_{C3.3V}}}{5\ \text{V}} = 4.47\ \text{A} \tag{2}$$

According to the graph displayed at Fig. 8, the expected efficiency of the 5 V DC-DC converter (Eff_{C5V}) for an output current of 4.47 A and an input voltage of 12 V is approximately 97.9%.

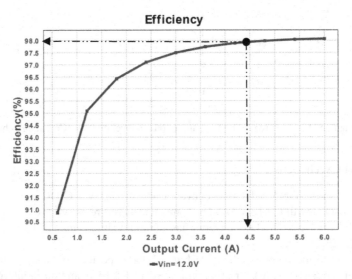

Fig. 8. Obtained efficiency regarding the output voltage level of 5 V (Cascade Architecture).

The output current required at the 12 V DC-DC converter for the 28 W requested by the LiDAR system and to power the 5 V DC-DC converter is:

$$Iout_{12Vconverter} = \frac{28\ \text{W} + \frac{5\ \text{W} + \frac{17\ \text{W}}{Eff_{C3.3V}}}{Eff_{C5V}}}{12\ \text{V}} \approx 4.24\ \text{A} \tag{3}$$

According to the graph displayed at Fig. 9, the expected efficiency of 12 V DC-DC converter (Eff_{C12V}) for an output current of 4.24 A and an input voltage of 65 V (worst case) is approximately 96.25%. Again, the efficiency is better for lower input voltages, namely 97.25% for an input voltage of 40 V and 98.7% for an input voltage of 15 V.

Therefore, the efficiency for the cascade architecture is:

$$\eta = \frac{50\ \text{W}}{\frac{28\ \text{W} + \frac{5 + \frac{17\ \text{W}}{Eff_{C3.3V}}}{Eff_{C5V}}}{Eff_{C12V}}} = 0.9467 = 94.67\% \tag{4}$$

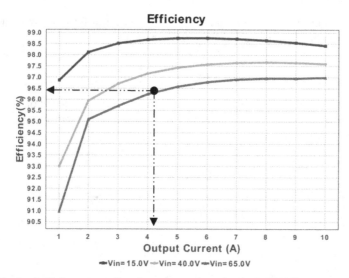

Fig. 9. Obtained efficiency regarding the output voltage level of 12 V (Cascade Architecture).

4.3 Hybrid Architecture

In this architecture, the 5 V DC-DC converter and the 3.3 V DC-DC converter are in parallel, cascaded with the 12 V DC-DC converter; therefore, they have the same input of 12 V, while the 12 V DC-DC converter has an input voltage range of 15 V to 65 V, as illustrated in Fig. 3 (a).

According to the graph displayed at Fig. 10, the expected efficiency of the 3.3 V DC-DC converter ($Eff_{H3.3V}$) for an output current of 5.15 A and an input voltage of 12 V is approximately 97.4%.

According to the graph displayed at Fig. 11, the expected efficiency of the 5 V DC-DC converter (Eff_{H5V}) for an output current of 1 A and an input voltage of 12 V is 97.4%.

The output current required at the 12 V DC-DC converter for the 28 W requested by the LiDAR system and to power the attached converters (5 V and 3.3 V) is:

$$Iout_{12Vconverter} = \frac{28\ W + \frac{5\ W}{Eff_{H5V}} + \frac{17\ W}{Eff_{H3.3V}}}{12\ V} \approx 4.22\ A \tag{5}$$

According to the graph displayed at Fig. 12, the expected efficiency of the 12 V DC-DC converter (Eff_{H12V}) for an output current of 4.22 A and an input voltage of 65 V (worst case) is approximately 96.25%. As in the former cases of the 3.3 V and 5 V DC-DC converters, the efficiency is better for lower input voltages, namely 97.25% for an input voltage of 40 V and 98.7% for an input voltage of 15 V.

Therefore, the efficiency for the hybrid architecture is:

$$\eta = \frac{50\ W}{\frac{28\ W + \frac{5\ W}{Eff_{H5V}} + \frac{17\ W}{Eff_{H3.3V}}}{Eff_{H12V}}} = 95.13\% \tag{6}$$

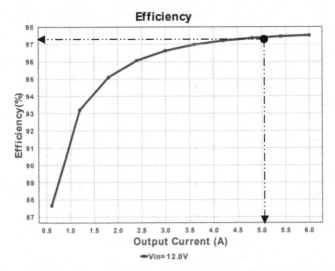

Fig. 10. Obtained efficiency regarding the output voltage level of 3.3 V (Hybrid Architecture).

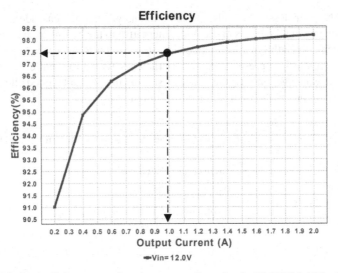

Fig. 11. Obtained efficiency regarding the output voltage level of 5 V (Hybrid Architecture).

4.4 Efficiency Comparison

The efficiency must be as high as possible to prevent unwanted power losses and optimize the LiDAR system performance. According to the detailed study performed focusing on the efficiency of the three possible architectures, and presented in the previous items, the obtained values of efficiency for each architecture are summarized in Table 3. By comparing the three architectures for the same conditions of operation, it was possible to identify that the hybrid architecture presents the best obtained efficiency, with a value

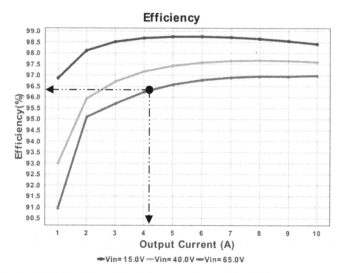

Fig. 12. Obtained efficiency regarding the output voltage level of 12 V (Hybrid Architecture).

of 95.13%. Nevertheless, it was also possible to analyze that the cascade architecture, with an efficiency of 94.67%, is also a good alternative to the hybrid architecture for the LiDAR system, since the obtained values are very similar to the cascade architecture. The parallel architecture presents the worst efficiency with a relevant difference when compared with the other architectures, demonstrating that it is not a good alternative for a LiDAR system due to the high-efficiency required.

Table 3. Comparison of estimated efficiencies among the architectures considered.

Architecture	Efficiency
Parallel	88.97%
Hybrid (a)	95.13%
Cascade	94.67%

5 Conclusions

This paper presented a comparison between different DC-DC converter architectures for a LiDAR system. The architectures under comparison were: three DC-DC converters linked in a single-input multiple-output (SIMO) structure; three DC-DC converters linked in cascade; and two DC-DC converters in a SIMO structure and in cascade with the third one. The DC-DC converters under comparison in the scope of this paper are based in the synchronous buck Switched-Mode Power Supply (SMPS) topology, and with the models LM5146-Q1, LM5116 and TPS548A20.

To evaluate and compare the performance of each DC-DC converter architecture, each converter was simulated with the WEBENCH® Power Designer software in order to determine its individual efficiency. The obtained results for the considered application show that the hybrid architecture has the highest efficiency (95.08%), considering the higher possible voltage at the input (65 V) and the maximum operating power of this particular LiDAR system. Taking into account that this DC-DC converter architecture accepts a wide range of input voltages and is able to provide multiple output voltages (12 V, 5 V, and 3.3 V), it is the most appropriate option considering the requisites of the presented LiDAR in terms of input voltage, output voltages, and different operating power levels. Only three of the possible variant architectures were used in the comparison to demonstrate the comparison method. The same methodology could have been used considering all possible hybrid variant architectures, the final best choice (best efficiency) will depend, in a general case, on the individual efficiencies exhibited by the DC-DC converters and the required power levels associated to each specified output voltage.

Acknowledgements. This work has been supported by national funds through FCT—Fundação para a Ciência e Tecnologia within the Project Scope: UIDB/00319/2020, and also European Structural and Investment Funds in the FEDER component, through the Operational Competitiveness and Internationalization Programme (COMPETE 2020) [Project nº 037902; Funding Reference: POCI-01-0247-FEDER-037902].

References

1. Manfreda, A., Ljubi, K., Groznik, A.: Autonomous vehicles in the smart city era: an empirical study of adoption factors important for millennials. Int. J. Inf. Manag. 102050 (2019). ISSN 0268-4012
2. Yoshioka, M., Suganuma, N., Yoneda, K., Aldibaja, M.: Real-time object classification for autonomous vehicle using LIDAR. In: 2017 International Conference on Intelligent Informatics and Biomedical Sciences (ICIIBMS), Okinawa, pp. 210–211 (2017). https://doi.org/10.1109/ICIIBMS.2017.8279696
3. Gerber, M., Ferreira, J.A., Hofsajer, I.W., Seliger, N.: High density packaging of the passive components in an automotive DC/DC converter. IEEE Trans. Power Electron. **20**(2), 268–275 (2005)
4. Van Breussegem, T., Wens, M., Redouté, J.M., Geukens, E., Geys, D., Steyaert, M.: A DMOS integrated 320 mW capacitive 12 V to 70 V DC/DC-converter for LIDAR applications. In: 2009 IEEE Energy Conversion Congress and Exposition, ECCE 2009, pp. 3865–3869 (2009)
5. Yang, J., Zhou, G., Yu, X., Zhu, W.: Design and implementation of power supply of high-power diode laser of LiDAR onboard UAV. In: 2011 International Symposium on Image and Data Fusion, Tengchong, Yunnan, China, pp. 1–4 (2011). https://doi.org/10.1109/ISIDF.2011.6024309.
6. Cui, X., Keller, C., Avestruz, A.: Cycle-by-cycle digital control of a multi-megahertz variable-frequency boost converter for automatic power control of LiDAR. In: 2019 IEEE Energy Conversion Congress and Exposition (ECCE), Baltimore, MD, USA, pp. 702–711 (2019). https://doi.org/10.1109/ECCE.2019.8913074
7. Bubovich, A.: The comparison of different types of DC-DC converters in terms of low-voltage implementation. In: 2017 5th IEEE Workshop on Advances in Information, Electronic and Electrical Engineering (AIEEE), Riga, Latvia, pp. 1–4 (2017). https://doi.org/10.1109/AIEEE.2017.8270560

8. Fu, J., Zhang, B., Qiu, D., Xiao, W.: A novel single-switch cascaded DC-DC converter of Boost and Buck-boost converters. In: 2014 16th European Conference on Power Electronics and Applications, Lappeenranta, Finland, pp. 1–9 (2014). https://doi.org/10.1109/EPE.2014.6910723

9. Chakraborty, S., Vu, H.N., Hasan, M.M., Tran, D.D., El Baghdadi, M., Hegazy, O.: DC-DC converter topologies for electric vehicles, plug-in hybrid electric vehicles and fast charging stations: State of the art and future trends. Energies 12(8), 1569 (2019)

10. Bhaskar Ranjana, M.S., Reddy, N.S., Pavan Kumar, R.K.: Non-isolated dual output hybrid DC-DC multilevel converter for photovoltaic applications. In: 2014 2nd International Conference on Devices, Circuits and Systems (ICDCS), Coimbatore, India, pp. 1–6 (2014). https://doi.org/10.1109/ICDCSyst.2014.6926197

11. Chang, R.C., Chen, W., Siao, C., Wu, H.: Low-complexity SIMO buck-boost DC-DC converter for gigascale systems. In: 2016 IEEE International Symposium on Circuits and Systems (ISCAS), Montreal, QC, pp. 614–617 (2016). https://doi.org/10.1109/ISCAS.2016.7527315

12. Kiguchi, R., Nishida, Y.: Boost DC-DC converter cascade system for high boost-rate application. In: 2018 International Conference on Smart Grid (icSmartGrid), Nagasaki, Japan, pp. 283–286 (2018). https://doi.org/10.1109/ISGWCP.2018.8634483

13. Texas Instruments: LM5146-Q1 100-V Synchronous Buck DC/DC Controller with Wide Duty Cycle Range. https://www.ti.com/product/LM5146-Q1

14. Texas Instruments: 6-100V Wide Vin, Current Mode Synchronous Buck Controller. https://www.ti.com/product/LM5116

15. Texas Instruments: 1.5 V to 20 V (4.5 V to 25 V Bias) Input, 15A SWIFT™ Synchronous Step-Down Converter. https://www.ti.com/product/TPS548A20

16. Sreedhar, J., Basavaraju, B.: Design and analysis of synchronous Buck converter for UPS application. In: 2016 2nd International Conference on Advances in Electrical, Electronics, Information, Communication and Bio-Informatics (AEEICB), Chennai, India, pp. 573–579 (2016). https://doi.org/10.1109/AEEICB.2016.7538356

17. Rashid, H.M.: DC-DC converters. In: Power Electronics Handbook, pp. 211–223. Academic Press (2003)

18. Iqbal, Z., Nasir, U., Rasheed, M.T., Munir, K.: A comparative analysis of synchronous buck, isolated buck and buck converter. In: 2015 IEEE 15th International Conference on Environment and Electrical Engineering (EEEIC), Rome, Italy, pp. 992–996 (2015). https://doi.org/10.1109/EEEIC.2015.7165299

19. Texas Instruments: WEBENCH® Power Designer. https://www.ti.com/design-resources/design-tools-simulation/webench-power-designer.html

Submodule Topologies and PWM Techniques Applied in Modular Multilevel Converters: Review and Analysis

Luis A. M. Barros[1]([✉]), Mohamed Tanta[1], António P. Martins[2], João L. Afonso[1], and José Gabriel Pinto[1]

[1] Centro ALGORITMI, University of Minho, Guimarães, Portugal
lbarros@dei.uminho.pt
[2] SYSTEC Research Center, University of Porto, Porto, Portugal

Abstract. Nowadays electrical energy presents itself as the most promising solution to satisfy the energy needs of smart cities. For electrical energy to be managed efficiently and sustainably, the use of power electronic converters is essential. The evolution of semiconductors, in terms of blocking voltages, conduted current and switching frequencies, led to the emergence of new topologies for more robust and compact power electronics converters with high-frequency galvanic isolation. However, for high and medium voltage applications, such as electric railways, wind turbines, solar photovoltaic, or energy distribution systems, the semiconductor blocking voltages are still below the values in demand. In order to solve this problem, modular multilevel converters (MMC) has been implemented, requiring more sophisticated control algorithms and complex pulse width modulation (PWM) techniques, such as space vector modulation.

In this paper, different topologies to be integrated into an MMC are presented, as well as PWM techniques for MMC, making a comparative analysis based on computer simulations of different PWM techniques developed in PSIM software. An MMC consisting of 4 full-bridge DC-AC power converters connected in series was considered as the study basis for the analysis of the PWM techniques.

Keywords: Level-shift PWM · Hybrid PWM · Modular multilevel converter · Phase-shift PWM · Phase disposition PWM · Pulse width modulation techniques

Nomenclature

v_{outx}	Instantaneous output voltage of the power submodule number x of the MMC.
v_{out}	Instantaneous voltage at the MMC output.
v_{load}	Instantaneous voltage on the load connected to the MMC.
v_{m_x}	Modulator waveform x.
v_{c_x}	Carrier waveform x.

J. L. Afonso et al. (Eds.): SESC 2020, LNICST 375, pp. 111–131, 2021.
https://doi.org/10.1007/978-3-030-73585-2_8

1   Introduction

Electricity is an essential asset for humanity, having a strong impact on the economic development of nations, on increasing the quality of life, and on sustainability. With the increase in population and the improvement in the quality of life, there is an incessant need for the availability of electricity, which makes it irreconcilable with existing energy sources [1]. Smart cities appear as an innovative concept in response to current needs, allowing the integration of renewable energy sources, energy storage locations, as well as the interface with a fully electrified transport network, as is the case with the rail system. That said, the importance of developing power electronics solutions that allow the gradual integration of new concepts and functionalities in smart cities becomes evident.

The topologies of two and three-level power electronics converters, similar to the half-bridge [2], full-bridge [2], or the basic cell structure of the neutral point clamping (NPC) [3], the T-type NPC (TNPC) [3], or the flying capacitor multilevel converter (FCMC) [4], are presented as the most used power electronics solutions in the most diverse applications. Its simple structure, with low voltages and power (up to 1 MVA), present interesting characteristics for various industrial applications, e.g., for driving motors, for interfacing renewable energy sources with the power grid, for charging of electric vehicles, among others [5]. However, for high/medium voltage or high power applications, such as large solar photovoltaic or wind turbine parks, and electric railway systems, these conventional solutions are no longer viable.

Analyzing the integration of high/medium voltage semiconductors (some kV) in conventional topologies, it is expected to have low switching frequencies. These factors are reflected in the high harmonic contents induced around the low switching frequency. In order to mitigate this problem, bulky passive filters are necessary, which increases the costs, as well as the losses of the converter. Additionally, in the conventional topologies, the DC-link voltage is applied to the converter output at high frequencies. Adopting these topologies to very high voltage applications, it is expected that the output voltage variations will be more accentuated, which would cause greater stress not only at the level of the semiconductors but also at the level of the galvanic isolation of any system connected to these terminals [6].

Modular multilevel converter (MMC) solution will allow significant mitigation of the aforementioned problems. MMC solutions are made up of different submodules connected in series and allows lower DC-link voltages. Additionally, it is possible to obtain different voltage levels depending on the number of submodules in the MMC. In this way, it is possible to obtain better sizing in terms of semiconductors and passive components. Nevertheless, and taking advantage of the concept of modularity, it is possible to easily adapt the solution to the voltage and power levels required by the final application. Barros et al. present in [7] different configurations for the MMC. Other MMC configurations are also presented in [8] and in [9].

Despite its applicability in high and medium voltage applications, the control of a modular system like MMC presents some complexity. In the literature, it is possible to find several scientific review papers that list the different pulse width modulation (PWM) techniques for MMC. However, the gap lies in comparing the performance of these different PWM techniques for MMC. The lack of this analysis makes it difficult to choose the best PWM technique to be implemented.

The purpose of this paper is to aid the implementation of MMC solution, presenting a comparative analysis for different PWM techniques for the MMC. In this sense, this paper is structured as follows: in Sect. 1, an introduction to the research topic is made, presenting some of the drawbacks of the conventional power converters; Sect. 2 presents the MMC concept, some of the applications as well as some power converters topologies to be integrated into the MMC solution; Sect. 3 is used to present some PWM techniques to be used in MMC application; Sect. 4 is dedicated to present the simulations results for different PWM techniques in a 4 full-bridge submodules connected in series developed in PSIM, as well as a comparative analyses are presented; Sect. 5 presents the main conclusions of this work.

2 Modular Multilevel Converter (MMC)

DC-AC power electronics converters, commonly referred to in the literature as "power electronics inverters", have the main purpose of synthesizing an AC output waveform, from a DC source on the input DC-link, using fully controlled semiconductors (e.g. isolated-gate bipolar transistor – IGBT, metal-oxide-semiconductor field-effect transistor – MOSFET, etc.). With the aid of control algorithms and the consequent activation of power semiconductors, it is possible to control the magnitude and the frequency of the synthesized waveform. The total control of these variables makes this type of converter ideal for motor drives, adjustable speed applications, active power filters, uninterruptible power supplies (UPS), and interface with AC power transmission/distribution systems (the public power grid) [2]. For the following explanations, an IGBT is used as a power switching device but the concepts is valid for other semiconductor devices.

DC-AC power electronics converters can be classified, according to the DC-link configuration, as a voltage source inverter (VSI) or current source inverter (CSI). This type of converters is characterized by discrete values at the output, requiring some precautions during their implementation in order to safeguard the integrity of the system. That is, these power converters are normally controlled by PWM, synthesizing a switched waveform with a switching frequency equal or multiple of the switching frequency of each device. This fact is characterized by the modulation technique used, normally having a carrier wave with the desired switching frequency. One of the most used techniques is the sinusoidal PWM (SPWM), where the modulating waveform is compared with a carrier, resulting in square pulses that will drive the IGBT with different actuation times, as used in [10]. Although the output does not have a sinusoidal waveform, as intended, the frequency of the fundamental component is close to the desired frequency. This fact takes some restrictions on its application as discussed in following.

In the case of VSI, it has high discrete values of dv/dt at the output. As a consequence, the load must be inductive in order to smooth the current waveform. If the load is capacitive, high current peaks will appear [2]. Thus, it is necessary to include an inductive filter between the load and the output of the VSI. On the other hand, in the case of CSI, it presents high discrete values of di/dt at the output. As a consequence, the load must be capacitive in order to smooth the voltage waveform [2]. If the load is inductive, high voltage peaks will appear. Thus, it becomes necessary to include a capacitive filter between the load and the CSI output [2].

Considering the integration of these conventional solutions in high voltage applications, in the case of VSI, it is necessary to increase the voltages in the DC-link. To be more specific, the DC-link voltage should be higher than the peak value of the power grid voltage at the point of common coupling for the correct operation of the DC-AC power converter [10].This requirement depends on the solutions available in the market, both at the level of semiconductor switching devices and at the level of high voltage capacitors. The available quantity of electronic components is one of the most important criteria for the development of power electronics converters in companies, and these specific characteristics have a limited stock. High voltage electronic components are not very common in large quantities at suppliers. On the other hand, increasing the operating voltage level on the DC-link will cause even greater dv/dt variations in the output of the VSI. Consequently, bulkier filters are needed.

MMC appears as a scalable solution for high voltage applications. This converter consists of several submodules that allow a better distribution of voltage stress and switching stress. The more the number of submodules, the higher the voltage level of MMC, and higher quality waveforms can be synthesized. Consequently, it is possible to reduce the output dv/dt variation, allowing the integration of smaller passive filters. Additionally, also enables the use and selection of more commercially available and robust power electronics components due to their wide applicability in the most diverse applications [7]. In other words, mature medium voltage technologies due to their strong applicability in other applications. Nevertheless, with the integration of MMC solutions, it is possible to increase the resulting output frequency without changing the switching frequency, taking advantage of the complementarity of the different submodules that constitute the MMC. As a consequence, the harmonic contents will be concentrated at high frequencies, being easily filtered [6]. Additionally, being a modular system, replacing a damaged submodule is easier as well as to adapt to the final application, adding submodules as much as need. Nevertheless, redundant protection mechanisms can be implemented [7].

Figure 1 illustrates some of the concepts mentioned in the previous paragraphs. By analyzing this figure, it is possible to conclude, both in the time domain and in the frequency domain, the impact of the number of levels that MMC can synthesize. It should be noted that, in the given example, the switching frequency of the fully controlled semiconductors in each power submodule is 20 times higher than the frequency of the fundamental component. Thus, it can be seen that the lower the number of levels of the output voltage, the more ample the harmonic spectrum will be, with a high concentration at low frequencies. In turn, with the increase in the number of voltage levels of the MMC, it can be seen that the harmonic content begins to tend towards high frequencies and with low amplitudes.

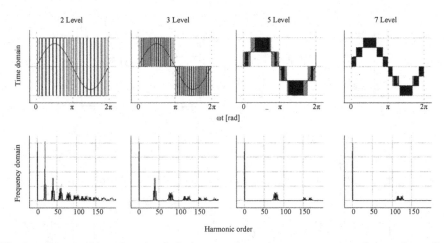

Fig. 1. Influence of the voltage levels that a power electronics converter can produce on the quality of the synthesized waveform.

2.1 MMC Submodule Topologies and Applications

Regarding the configurations of submodules that constitute the MMC, different approaches have been explored. Perez et al. present in [8] different configuration approaches for different submodules. Furthermore, they also present different topologies of power electronics converters that can be integrated into the different MMC power submodules. Similarly, Feng et al. present in [9] different MMC solutions that reputable companies related to railway traction, such as ABB, Alstom, Siemens, and Bombardier have been adopting in the implemented concepts. Barros et al. present in [7] some trends in the MMC configurations.

Figure 2 shows examples of topologies of power electronics converters to be used as an MMC submodule. A more detailed analysis of the functionalities of each topology, highlighting advantages and challenges of implementation, is presented in the following items.

2.1.1 Half-Bridge Submodule

The half-bridge power electronics converter with split dc-link, represented in Fig. 2 (a), has a simple structure among the other power converter submodules, consisting of only two capacitors on the DC-link and two IGBT devices connected in series. The midpoint of the capacitors in the DC-link is used as a reference for the output voltage, v_{out}, of the converter. Consequently, for a maximum output voltage of V_{DC_link}, the DC-link of the converter will have to withstand twice that voltage with each capacitor having a DC voltage of V_{DC_link} at its terminals. The operation principle as well as the permitted operating states are presented in [2].

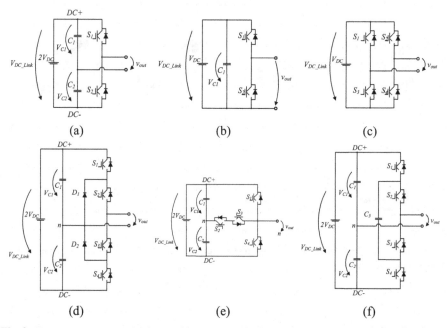

Fig. 2. Power converter topologies used in MMC applications: (a) Half-bridge with split DC-link; (b) Half-bridge; (c) Full-bridge; (d) NPC; (e) T-NPC; (f) FCMC.

Another approach of the half-bridge converter is shown in Fig. 2 (b), the difference here is being used a single capacitor on the DC-link. The operation principle as well as the permitted operating states are presented in [2]. Although it only can create an output voltage with a DC component, the half-bridge topology with a single capacitor on the DC-link, is one of the most used typologies for MMC applications. In order to allow the synthesis of an AC waveform without a continuous component, it is necessary to have a complementary submodule. Thus, a submodule is responsible for synthesizing the positive semi-cycle, the second module being responsible for synthesizing the negative semi-cycle. This submodule only works in the first two quadrants (positive voltage only and bidirectional current). It has the simplest structure among the other MMC submodules.

An example of an MMC based on half-bridge submodules with a single DC-link acting as a railway power conditioner (RPC) in railway applications is presented by Tanta et al. in [25]. A 50 kVA three-phase prototype with 10 submodules per phase is presented in [26], where Moranchel et al. presents a comparison of different modulation techniques.

The biggest implementation challenges based on this topology lies essentially in balancing the DC-link voltage of each submodule. In this context, deadbeat predictive current control methodology to reduce the circulating currents and balancing the submodule voltages in the MMC is presented in [11]. Moreover, each submodule will contribute to a portion of the total power, which causes currents circulating in the submodules. Additionally, and in order to generate an AC component, the duplication of submodules, as well as capacitors, is expected, which will inflate the cost of implementation as well as the complexity of the system [22].

2.1.2   Full-Bridge Submodule

The full-bridge DC-AC power converter, commonly known as H-bridge converter, consists of a DC-link and two arms, each arm being composed of two IGBT devices. The output of the converter is at the midpoints of each arm, as shown in Fig. 2 (c). Each IGBT device that constitutes a converter arm works in a complementary way. The operation principle as well as the permitted operating states are presented in [2]. Due to the semiconductor arrangement, two separate arms in parallel, and with a unipolar modulation technique, it is possible to obtain a resulting frequency at the power converter output voltage with twice the value of the switching frequency imposed on the IGBT devices. This feature is especially important in practical development, making it possible to reduce the passive output filters.

In relation to the practical implementation of this topology, it can be seen that for the same level of output voltage, the voltage on the DC-link is half the level obtained in solutions with a split DC-link. Additionally, due to the existence of two IGBT devices referenced to the same potential, it allows the development of cheaper and more robust driver circuits. In fact, one interesting topology consists of connecting in parallel different full-bridges, sharing the same DC-link. The output of each submodule is connected to a low-frequency power transformer, as presented in [12], with the secondary windings connected in series. The power transformers can have different turns ratio, which gives a variety of output levels even with few submodules. Note that this solution has several IGBT devices connected to the same electrical potential, which allows the implementation of more robust and cheaper gate driver circuits. In order to synthesize all voltage levels, this solution requires a control algorithm capable of detecting the combination of states that allows the greatest reduction in the output error of the generated voltage. One solution is to implement space vector modulation (SVM). However, the lower the switching frequency of the submodules, the greater the transferred power. In [12], the authors state that the module with less frequency can transfer up to 80% of the total power. Thus, it is concluded that this type of solution would not be the most suitable for a modular implementation.

In [9], different implemented MMC solutions composed of full-bridge power converters are presented, highlighting the combination of these converters with solid-state transformers (SST).

2.1.3 Neutral Point Clamped Submodule

The neutral point clamped (NPC) power converter, illustrated in Fig. 2(d), is essentially made up of an arm with capacitors, with two sets of capacitors connected in series, which constitute the DC-link, and another arm made up of power semiconductors.

Regarding the operating mode, each IGBT device that constitutes a converter arm is activated in a complementary way, in order to control the output voltage. This topology has the versatility of obtaining two or three voltage levels at the output, v_{out}, depending on the unipolar or bipolar PWM technique adopted. Due to the greater predominance of semiconductors in this topology, being the only passive components in the DC-link, only the conduction and switching losses caused by the diodes and the IGBT devices are considered [3].

Regarding the complexity of implementation and system performance, with the increase in voltage levels generated by the converter, the number of capacitors on the DC-link also increases. Despite the similarity of the components, there will always be voltage imbalances in different capacitors caused by system impedances. In addition, the PWM technique adopted may intensify this imbalance [3]. Due to this intrinsic characteristic, the voltage imbalance on the DC-link presents itself as the main disadvantage of this topology. Different balancing techniques have been developed, namely active control solutions [3, 13, 14], active auxiliary circuits [3, 15], or passive auxiliary circuits [3, 16]. With regard to balancing techniques based on control algorithms, they make the system more complex. In turn, solutions with auxiliary circuits result in greater energy losses, greater complexity of implementation, and higher costs due to peripherals and additional components.

2.1.4 T-type Neutral Point Clamping Submodule

The T-type NPC (TNPC), also known as a neutral point pilot (NPP) [17, 18], has a structure similar to the half-bridge converter with a split DC-link. In addition, T-NPC contains fully controlled bidirectional switches between the ac terminals. The fully controlled bidirectional switches essentially are two anti-series IGBT devices, as presented in Fig. 2 (e). Regarding the DC-link, and similar to the NPC, T-NPC consists of two sets of capacitors connected in series. The upper bus is connected between the positive terminal and the neutral point, and the lower bus is connected between the neutral point and the negative terminal. The midpoint of the capacitors in the DC-link represents not only the neutral point of the circuit but also the reference point for output voltage. Despite the T-NPC is considered as a multilevel converter, the authors of this paper could not find relevant research implemented with the T-NPC as MMC. However, being to be associated in review papers as a potential submodule of an MMC [18].

2.1.5 Flying Capacitor Multilevel Converter Submodule

The flying capacitor multilevel converter (FCMC), introduced by Meynard and Foch, has a similar structure to the NPC, replacing the pair of diodes with a capacitor, as shown in Fig. 2 (f) [19]. Consequently, the operating state in which the two IGBT devices in the middle are conducting is no longer allowed, as it would cause a short circuit to the capacitor [4]. Consequently, the blocking voltage for each IGBT device should be at least V_{DC}. For the correct operation of the power converter, the flying capacitor, C3, must have a voltage at its terminals equal to the voltage of each DC-link capacitor [20]. Regarding the DC-link, and similar to the NPC, it consists of two sets of capacitors connected in series. The upper bus is connected between the positive terminal and the neutral point, and the lower bus is connected between the neutral point and the negative terminal. The midpoint of the capacitors in the DC-link represents not only the neutral point of the circuit but also the reference point of the output voltage.

Due to the predominance of capacitors in this topology, inrush currents at the start of operation may occur. For this reason, and in order to mitigate large current transients and safeguard the integrity of the system, it is necessary to implement pre-charge circuits [3]. A common practice in order to minimize inrush currents is to insert a series resistor between the capacitors and a DC source during the pre-charge process. Once the minimum voltage value of the capacitors has been obtained, a switch in parallel with a resistor is activated in order to bypass and, consequently, to minimize the energy losses, caused by the resistor, during the system's steady-state operating regime.

Continuing with the topology analysis, it can be seen that the capacitors will play an important role in the system. In fact, the amount of energy stored in the capacitors is directly related to the voltage ripple in each capacitor and, consequently, the performance of the converter. In experimental tests carried out in [21], the authors found that it would be possible to decrease the capacitors' capacity by increasing the switching frequency. However, this approach leads to problems with electromagnetic compatibility. Nevertheless, the reliability of the capacitors, namely electrolytic ones, in power electronics circuits is a major concern. In addition, it is important to consider the space occupied by electrolytic capacitors compared to the space occupied by the diodes of, for example, the NPC topology [3].

2.1.6 MMC Submodules Topologies Comparison

In this item, a comparative analysis of the different topologies addressed is carried out. As it was possible to verify, the different topologies present several modes of operation, restrictions, and requirements for their correct functionality. Table 1 presents some comparative topics between the topologies. It should be noted that the analysis is done from the application point of view, with a requirement for a given minimum voltage value at the converter output, which is the same in all, being necessary to size or adapt the remaining system variables, as is the case of the voltage level on the DC-link. In [22] a more detailed study of more topologies is presented.

By analyzing Table 1, it can be seen that the existence of additional IGBT devices allows the creation of more voltage levels at the converter output. In the case of the half-bridge power converter, since it only has two IGBT devices, it can only generate two voltage levels. In turn, the remaining topologies take advantage of 4 IGBT functionality to be able to generate 3 voltage levels. Nevertheless, the constitution of a single capacitor on the DC-link is simpler in the full-bridge topology, providing robust voltage regulation and simple DC-link control algorithms. Despite the similarity in the DC-link level of the half-point topology with a single bus, for AC applications, it is necessary to insert an additional module responsible for the negative semi-cycle. This fact makes it necessary to duplicate the electronic components used, as well as greater complexity in terms of control algorithms. Additionally, it is possible to conclude that topologies with split DC-link require double the voltage of the DC-link in order to generate the same output voltage level as the full-bridge can.

Table 1. Comparative table of power converter topologies.

	Half-Bridge (i)	Half-Bridge (ii)	Full-Bridge	NPC	TNPC	FCMC
Number of output levels	2	2	3	3	3	3
Number of DC-link capacitors	2	1	1	2	2	2
Number of IGBT	2	2	4	4	4	4
Number of diode	0	0	0	2	0	0
Unipolar operation	No	No	Yes	Yes	Yes	Yes
Max voltage block	V_{DC}	V_{DC}	V_{DC}	V_{DC}	$2V_{DC}$	V_{DC}
DC-link voltage	$2V_{DC}$	$2V_{DC}$	V_{DC}	$2V_{DC}$	$2V_{DC}$	$2V_{DC}$
Output frequency	f_s	f_s	$2f_s$	f_s	f_s	f_s
Cell design complexity	*	**	*	**	**	**
Cell control complexity	**	**	*	**	**	***

*Simple; **Moderate; ***Complex

Another interesting fact for practical implementation resides in the resulting frequency of the output voltage from the topologies. Since the full-bridge is the only one that has two arms with IGBT devices in parallel, it can provide a resulting frequency from the output wave with twice the value of the frequency used for switching. This characteristic is quite advantageous in terms of practical implementation and it can result in a reduction in terms of passive filters which consequently translates into the reduction of losses essentially in the core of the magnetic components.

3 PWM Technique for MMC Application

In the literature, there are several PWM techniques for controlling different power electronics converters with 2, 3 or more voltages levels. Regarding the MMC, the PWM algorithms are more complex, with different stages for each power submodule in order to synthesize multilevel waveform. Thus, this topic is dedicated to study the PWM techniques for MMC.

3.1 Level-Shift PWM Technique

The level-shift PWM techniques were initially proposed for the control of MMC based on topologies such as NPC or FCMC. In turn, when these modulation techniques were implemented in MMC solutions consisting of full-bridge submodules, it caused power imbalances in different submodules. In addition to the imbalance in the level of operation functioning between submodules, this fact also causes circulating currents between submodules as well as the injection of harmonics into the power grid. However, different changes to the PWM techniques have been made in order to provide a greater balance of power in each submodule [23].

Within the level of modulation techniques deviation, one can enumerate the phase disposition (PD), phase opposition disposition (POD), and the alternate phase opposition disposition (APOD). In general, carrier waveforms have the same amplitude and frequency but are arranged vertically with different average values (offsets). The carrier waveforms of the PD modulation technique have the same phase angle, as represented in Fig. 3(a). The carrier waveforms of the POD modulation technique have symmetry in relation to the zero reference line, with the carrier waveforms lower than this value in phase with each other but 180° out of phase in relation to the upper carrier waves, as can be seen in Fig. 3(c). Finally, the carrier waveforms of the APOD modulation technique are alternately 180° out of phase, as shown in Fig. 3(d). Adaptive modulation technique to balance the interface powers in each submodule when PD modulation technique is presented in Fig. 3(b). For this balanced technique, where can be used with different PWM techniques, a carrier permutation is necessary [24].

Another variant of the level-shift modulation technique consists of overlapping carrier waveforms, being designated in the nomenclature as carrier overlapping (CO). In this modulation technique, the carrier waveforms share the same value of frequency and phase angle, varying the value of amplitude and offset. The overlap value is defined as half the amplitude of the different adjacent carrier waveforms, as shown in Fig. 3(e). The correct sizing of the modulation index for the CO modulation technique allows a better performance at the level of harmonic content when compared with the aforementioned modulation techniques. This fact is accentuated with modulation indices below 70%. Due to the overlap of the carrier waveforms, the reference wave intercepts more times minimizing the dispersion of the harmonic content. However, for a modulation index above 80%, the harmonic distortion begins to approach between the different techniques [23]. Despite its existence in the literature, there is no scientific content that portrays its practical implementation.

The most common practice regarding the implementation of level-shift modulation techniques, consists only of the variation in the vertical axis, amplitude, and average value, always keeping the frequency of the carrier waves constant. However, with variable frequency (VF), it is possible to obtain new features. By varying the frequency of the carrier waves, it is possible to change the harmonic spectrum of the output waveform through harmonic cancellation. Thus, in MMC applications, when the output current reaches peak values, the voltage of the capacitors in the peripheral submodules tends to fluctuate [23]. In order to mitigate this phenomenon, a higher frequency in the peripheral carrier waves of the levels was considered as represented in Fig. 3(f). With the increase of the frequency in the peripheral submodules, it is possible to decrease the voltage ripple of the existing waveform in the DC-link capacitors and, consequently, improve the quality of the output waveform. In turn, another variant of this control, the VF2, was initially created in order to equalize the losses along with the NPC topology [23]. Since the IGBT devices positioned on the periphery of the converter switch at a higher frequency, the frequency of the intermediate carrier waves was increased in order to equalize the transitions, as illustrated in Fig. 3(g).

3.2   Phase-Shift Carrier PWM Technique

The phase-shifted modulation technique is presented as the pioneering modulation technique for the control of multiple modules of power electronics converters. In this modulation technique, all carrier waveforms share the same amplitude, frequency, and offset value, varying the phase angle between them. This method is exemplified in Fig. 3(h), where the carrier waveforms are displaced $2\pi/N$ between them. N represents the number of submodules in one MMC arm. This modulation technique minimizes the voltage ripple in the DC-link capacitors, as well as eliminates some harmonic contents in the output voltage waveform. From the point of view of practical implementation for MMC solutions, this modulation technique provides a greater balance in terms of power of the different submodules. However, it has no sensitivity in the fluctuation of the DC-link voltages, unlike the VF technique [23].

3.3 Hybrid Carrier PWM Technique

Hybrid modulation methods take advantage of the best features of previous techniques. However, due to the restrictions of this modulation technique, the number of levels, M, of the converter output has to be odd. Within the hybrid modulation techniques, it is possible to highlight the phase-shift disposition (PSD), carrier overlapping disposition (COD), carrier overlapping phase disposition (CO-PD), Carrier overlapping opposition disposition (CO-POD), Carrier overlap-alternate phase opposition disposition (CO-APOD).

The PSD technique arises from the combination of PSC and PD modulation techniques to control an MMC with M voltage levels. This is characterized by the existence of $(M-1)/2$ carrier waves with values above the zero reference value and with the disposition of PSC. For the remaining $(M-1)/2$ carrier waves with values below the zero reference value, these show symmetry with the upper carrier waves, as can be seen in Fig. 3(i). This modulation method is characterized by the existence of harmonic content concentrated in the value $f_C (M-1)/2$, where f_C represents the carrier frequency. Studies have shown that whenever the modulation index of this control technique exceeds 70%, the system presents less switching losses than PD, POD, APOD, CO, PSC, and COD [23].

The PWM carrier overlapping disposition (COD) modulation technique results from the combination of the CO technique with the level-shift technique. Considering a converter with M levels, it is necessary to implement two groups of $(M-1)/2$ carrier waves, configured as shown in Fig. 3(j). The combination of the carried waves with different lags results in the modulation techniques seen below. Maintaining the phase angle between the carrier waves, the carrier overlapping-phase device (CO-PD) modulation technique is obtained, as represented in Fig. 3(j). In turn, by changing the phase angle of the group of carrier waves that are below the x-axis, as shown in Fig. 3(k), the carrier overlapping-opposition disposition (CO-POD) modulation technique is obtained. In turn, maintaining the 180° offset in relation to the adjacent waves, the carrier overlapping alternate phase opposition disposition (CO-APOD) modulation technique is obtained, as can be seen in Fig. 3(l).

It should be mentioned that the y-axis represents the amplitude of the carrier waves and the modulating waves. For digital programming, the amplitudes of the carrier waves would represent the maximum value of the counter register used in the PWM peripheral. Regarding the modulating wave, this represents the control variable originated in an output current control algorithm, for example, requiring its vertical displacement to be positive values.

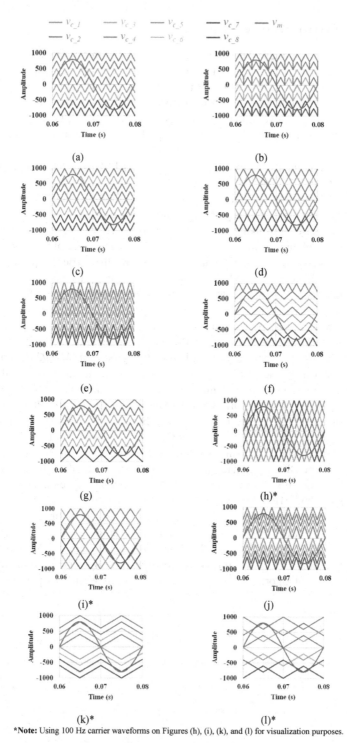

*Note: Using 100 Hz carrier waveforms on Figures (h), (i), (k), and (l) for visualization purposes.

Fig. 3. PWM techniques: (a) PD; (b) Balanced PD (c) POD; (d) APOD; (e) CO; (f); VF (g) VF2; (h) PSC; (i) PSD; (j) CO-PD; (k) CO-POD; (l) CO-APOD.

4 Simulation Results

The topology adopted for the object of study is composed of four power submodules connected in series that integrate a full-bridge in each one. A DC source is connected to the DC-link of each submodule in order to maintain a stable DC-link voltage, as shown in Fig. 4(a). The circuit used to generate the PWM signal are represented in Fig. 4(b). The parameters used for the simulation developed in PSIM are shown in Table 2. That said, in the following topics description and analysis of different modulation techniques are made, as well as a practical comparison, based on computer simulation results, of the PWM strategy with the best results for the adopted topology. For the simulation, two switching frequency were considered, one of 500 Hz for visualization and comparative analysis reasons, and another one of 20 kHz for a more realistic real application approach.

Table 2. Main parameters of the computer simulation for different PWM techniques

Description	Variable	Value
Fundamental frequency	F	50 Hz
Modulation Index	m_a	0.8
Output inductance	L_s	5 mH
Internal resistance	R_s	2 Ω
Load	R_{load}	10 Ω
DC-link voltage	V_{DC}	175 V
Switching frequency	f_s	500 Hz and 20 kHz

Figure 5 shows the voltage waveforms at the output of the MMC, v_{out}, and in the load, v_{load}. Analyzing the waveforms of Fig. 5(a) representative of the PD PWM technique, it can be seen that it presents an asymmetry in relation to the x-axis. However, using the technique of permutation of carrier waves to balance the interface power of each submodule, represented in Fig. 5(b), it can be seen that it does not have a significant impact on the waveforms. In turn, when the POD technique is implemented, Fig. 5(c), it imposes to be symmetric in relation to the x-axis. Analyzing the remaining level-shift techniques, the APOD represented in Fig. 5(d), the CO represented in Fig. 5(e), VF in Fig. 5(f), and VF2 in Fig. 5(g), it can be seen that all of them have an asymmetry of operation. It should be noted that with the CO technique, there is a greater number of v_{out} waveform transitions due to the overlap of the carried waves, which results in a greater number of interceptions with the modulating wave. Regarding the PSC technique, Fig. 5(h), and since the carrier waves are all arranged horizontally, it can be seen that v_{out} tends to obtain a sinusoidal waveform even at low frequencies. Additionally, there

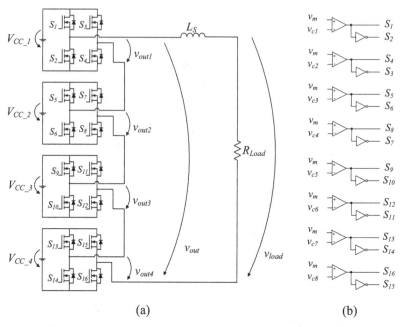

(a)   (b)

Fig. 4. Electrical schematic of: (a) MMC composed of 4 full-bridge submodules; (b) Control circuit to generate the PWM signals.

is a total symmetry in its operation. In turn, with the PSD, represented in Fig. 5(i), and since the carrier waves are grouped into two groups, the intersections will be smaller in each half-cycle, resulting in a slight degradation of the v_{load} waveform. However, it is still possible to verify the existence of an operation symmetry. Finally, the hybrid techniques CO-PD, CO-POD, and CO-APOD, illustrated in Fig. 5(j), Fig. 5(k), and Fig. 5(l), respectively, present the best characteristics of each original technique: a greater number of v_{out} transitions due to the overlap of the carrier waves that results in a v_{load} waveform close to the sinusoidal even at low frequencies; and a symmetry of operation originating from the POD technique.

For practical implementation, the PSC PWM technique presents as the best PWM technique for controlling an MMC. In addition to presenting good results even at low switching frequencies, it is easy to implement in a dedicated digital signal controller (DSC) platform for power electronics converters. Considering the Texas Instruments DSC C2000 real-time controllers, these have specific registers for horizontal displacement, which facilitates the implementation. In turn, the implementation of vertical displacement in digital control platforms presents greater difficulty, requiring the segmentation of the modulating waveform, v_m, within a certain interval (imposed by the period counter register) and its restructuring in order to maintain the same points of comparison (imposing symmetries in order to maintain the same pulse sequence). Nevertheless, the PWM techniques with asymmetry operation will equally impose an asymmetry operation of the control system when implementing an output current control algorithm and, in some situations, the need for different gains for control of the output signal at the maximum and minimum values. In other words, the PWM techniques with symmetric operation will facilitate the adjustment of gains of the control algorithm used.

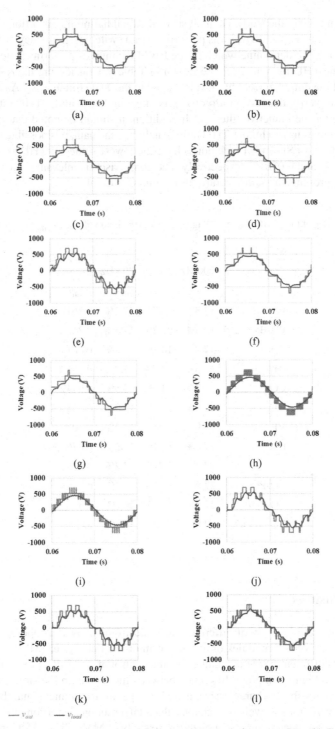

Fig. 5. Output voltage and load voltage using the PWM technique: (a) PD; (b) Balanced PD (c) POD; (d) APOD; (e) CO; (f); VF (g) VF2; (h) PSC; (i) PSD; (j) CO-PD; (k) CO-POD; (l) CO-APOD.

Proceeding with the study, an analysis of the total harmonic distortion (THD) ratio of the waveforms v_{out} and v_{load} was performed for a switching frequency of 500 Hz and 20 kHz in order to determine which of the PWM techniques performs better. The final results of the THD ratio analysis are shown in Table 3. Considering the resistive load, the THD of the output current has a value equal to the THD of the v_{load}. As expected, it is possible to verify that THD ratio of v_{out} is much higher than the THD ratio of v_{load}, measured after the inductive filter, L_s. In addition, it should be noted that overlapping PWM techniques have a higher THD ratio. Analyzing the data, it is possible to conclude that the PSC and PSD PWM techniques have the lowest values of THD ratio for v_{load}, with 0.0768% and 0.0782% respectively. The comparison is relevant since the THD of v_{load} also represents the same feature for the output current.

Table 3. THD$_\%$ of the MMC output voltage (v_{out}) and the load voltage (v_{load}) for different PWM techniques.

	500 Hz		20 kHz	
	v_{out}	v_{load}	v_{out}	v_{load}
PD	17.3%	9.67%	17.2%	0.261%
PD-Balanced	17.3%	9.42%	17.3%	0.343%
POD	17.3%	10.7%	17.2%	0.261%
APOD	17.3%	10.7%	17.2%	0.261%
CO	26.3%	14.9%	24.6%	0.944%
VF2	17.6%	9.93%	17.3%	0.303%
VF	16.9%	11%	17.2%	0.500%
PSC	17.4%	2.71%	17.2%	0.0768%
PSD	17.4%	2.71%	17.2%	0.0782%
CO-PD	27%	16.1%	26.2%	6.03%
CO-POD	27.5%	17.3%	26.2%	6.03%
CO-APOD	20.8%	12.2%	18.8%	6.02%

5 Conclusions

This paper presents a study of different power electronic converter topologies that can integrate the power submodules of a modular multilevel converter (MMC). The paper also presents different PWM techniques for the control of an MMC, presenting simulation results to highlighting the difference between them. Within the simulation results obtained, it is possible to characterize the mode of operation (symmetry, number of levels and voltage variation), as well as to analyze the total harmonic distortion ratio originated in each modulation technique for a certain case study of MMC. The MMC in this paper is composed by 4 full-bridge submodules connected in series, with an inductive filter and a resistive load.

Based on this study, it was possible to verify the PWM techniques whose carrier waveforms are symmetrical in relation to the x-axis, allowing to synthesize symmetrical output waveforms. In addition, it was found that modulation techniques with overlapping carrier waveforms result in a greater number of interceptions with the modulating waveform and, consequently, more transitions in the MMC output waveform. Regarding the THD ratio, it was found that the phase-shift carrier (PSC) and the phase-shift disposition (PSD) PWM techniques allow to synthesize an output waveform with low THD ratio. Consequently, this allows to reduce the passive filters used. Additionally, the PSC is presented as a technique that is easy to implement in a digital signal controller platform, where there are dedicated registers for the horizontal offset of the carrier waveforms. Thus, and for the MMC applications, more specifically made up of full-bridges submodules, the PSC modulation technique is recommended both for its ease of implementation and for the results obtained.

Due to the advantages they present in relation to conventional topologies, MMC will contribute to greater sustainability of electrical systems, the most promising energy source of future smart cities.

Acknowledgements. This work has been supported by FCT – Fundação para a Ciência e Tecnologia with-in the Project Scope: UIDB/00319/2020. This work has been supported by the FCT Project QUALITY4POWER PTDC/EEI-EEE/28813/2017. Mr. Luis A. M. Barros is supported by the doctoral scholarship PD/BD/143006/2018 granted by the Portuguese FCT foundation. Dr. Mohamed Tanta was supported by FCT PhD grant with a reference PD/BD/127815/2016.

References

1. Barros, L.A.M.: Desenvolvimento de um microinversor com armazenamento local de energia para aplicações solares fotovoltaicas. M.Sc. thesis (2016)
2. Rashid, M.H.: Power Electronics Handbook. Butterworth-Heinemann, Oxford (2017). ISBN 978-0-12-382036-5
3. Maswood, A.I., Tafti, H.D.: Advanced Multilevel Converters and Applications in Grid Integration. Wiley, Hoboken (2018). ISBN 978-1-119-47589-7
4. Brenna, M., Foiadelli, F., Zaninelli, D.: Electrical Railway Transportation Systems, vol. 67. Wiley, Hoboken (2018). ISBN 978-1-119-38680-3
5. Steimel, A.: Electric Traction - Motive Power and Energy Supply: Basics and Practical Experience. Oldenbourg Industrieverlag, München (2008). ISBN 978-3-8356-3132-8
6. Sharifabadi, K., Harnefors, L., Nee, H.-P., Norrga, S., Teodorescu, R.: Design, Control, and Application of Modular Multilevel Converters for HVDC Transmission Systems. Wiley, Hoboken (2016). ISBN 978-1-118-85156-2
7. Barros, L.A.M., Tanta, M., Martins, A.P., Afonso, J.L., Pinto, J.G.: Opportunities and challenges of power electronics systems in future railway electrification. In: CPE - POWERENG 2020: 14th International Conference on Compatibility, Power Electronics and Power Engineering. IEEE (2020). https://doi.org/10.1109/CPE-POWERENG48600.2020.9161695

8. Perez, M.A., Bernet, S., Rodriguez, J., Kouro, S., Lizana, R.: Circuit topologies, modeling, control schemes, and applications of modular multilevel converters. IEEE Trans. Power Electron. 30(1), 4–17 (2014). https://doi.org/10.1109/TPEL.2014.2310127

9. Feng, J., Chu, W., Zhang, Z., Zhu, Z.: Power electronic transformer-based railway traction systems: Challenges and opportunities. IEEE J. Emerg. Sel. Topics Power Electron. 5(3), 1237–1253 (2017). https://doi.org/10.1109/JESTPE.2017.2685464

10. Barros, L.A.M., Tanta, M., Sousa, T.J.C., Afonso, J.L., Pinto, J.G.: New multifunctional isolated microinverter with integrated energy storage system for pv applications. Energies 13(15), 4016 (2020). https://doi.org/10.3390/en13154016

11. Tanta, M., Pinto, J., Monteiro, V., Martins, A.P., Carvalho, A.S., Afonso, J.L.: Deadbeat predictive current control for circulating currents reduction in a modular multilevel converter based rail power conditioner. Appl. Sci. 10(5), 1849 (2020). https://doi.org/10.3390/app100 51849

12. Latran, M.B., Teke, A.: Investigation of multilevel multifunctional grid connected inverter topologies and control strategies used in photovoltaic systems. Renew. Sustain. Energy Rev. 42, 361–376 (2015). https://doi.org/10.1016/j.rser.2014.10.030

13. Ye, Z., Xu, Y., Wu, X., Tan, G., Deng, X., Wang, Z.: A simplified PWM strategy for a neutral-point-clamped (NPC) three-level converter with unbalanced DC links. IEEE Trans. Power Electron. 31(4), 3227–3238 (2015). https://doi.org/10.1109/TPEL.2015.2446501

14. Pou, J., et al.: Fast-processing modulation strategy for the neutral-point-clamped converter with total elimination of low-frequency voltage oscillations in the neutral point. IEEE Trans. Ind. Electron. 54(4), 2288–2294 (2007). https://doi.org/10.1109/TIE.2007.894788

15. Filba-Martinez, A., Busquets-Monge, S., Bordonau, J.: Modulation and capacitor voltage balancing control of multilevel NPC dual active bridge DC-DC converters. IEEE Trans. Ind. Electron. 67(4), 2499–2510 (2019). https://doi.org/10.1109/TIE.2019.2910035

16. Stala, R.: Application of balancing circuit for DC-link voltages balance in a single-phase diode-clamped inverter with two three-level legs. IEEE Trans. Ind. Electron. 58(9), 4185–4195 (2010). https://doi.org/10.1109/TIE.2010.2093477

17. Akagi, H.: Multilevel converters: Fundamental circuits and systems. Proc. IEEE 105(11), 2048–2065 (2017). https://doi.org/10.1109/JPROC.2017.2682105

18. Dekka, A., Wu, B., Fuentes, R.L., Perez, M., Zargari, N.R.: Evolution of topologies, modeling, control schemes, and applications of modular multilevel converters. IEEE J. Emerg. Sel. Topics Power Electron. 5(4), 1631–1656 (2017). https://doi.org/10.1109/JESTPE.2017.274 2938. ISSN 2168-6777

19. Meynard, T., Foch, H.: Multi-level conversion: high voltage choppers and voltage-source inverters. In: PESC 1992 Record. 23rd Annual IEEE Power Electronics Specialists Conference, pp. 397–403 (1992). https://doi.org/10.1109/PESC.1992.254717

20. Gonzalez, S.A., Verne, S.A., Valla, M.I.: Multilevel Converters for Industrial Applications. CRC Press, Boco Raton (2016). https://doi.org/10.1109/MIE.2014.2299500. ISSN 1932-4529

21. Fazel, S.S., Bernet, S., Krug, D., Jalili, K.: Design and comparison of 4-kV neutral-point-clamped, flying-capacitor, and series-connected H-bridge multilevel converters. IEEE Trans. Ind. Appl. 43(4), 1032–1040 (2007). https://doi.org/10.1109/TIA.2007.900476

22. Nami, A., Liang, J., Dijkhuizen, F., Demetriades, G.D.: Modular multilevel converters for HVDC applications: review on converter cells and functionalities. IEEE Trans. Power Electron. **30**(1), 18–36 (2015). https://doi.org/10.1109/TPEL.2014.2327641. ISSN 0885-8993
23. Antonio-Ferreira, A., Collados-Rodriguez, C., Gomis-Bellmunt, O.: Modulation techniques applied to medium voltage modular multilevel converters for renewable energy integration: a review. Electr. Power Syst. Res. **155**, 21–39 (2018). https://doi.org/10.1016/j.epsr.2017.08.015
24. Sochor, P., Akagi, H.: Theoretical and experimental comparison between phase-shifted PWM and level-shifted PWM in a modular multilevel SDBC inverter for utility-scale photovoltaic applications. IEEE Trans. Ind. Appl. **53**(5), 4695–4707 (2017). https://doi.org/10.1109/TIA.2017.2704539. ISSN 0093-9994
25. Tanta, M., et al.: Experimental validation of a reduced-scale rail power conditioner based on modular multilevel converter for AC railway power grids. Energies **14**(2), 484 (2021). https://doi.org/10.3390/en14020484
26. Moranchel, M., Huerta, F., Sanz, I., Bueno, E., Rodriguez, F.J.: A comparison of modulation techniques for modular multilevel converters. Energies **9**(12), 1091 (2016). https://doi.org/10.3390/en9121091

Mitigation of Low-Frequency Oscillations by Tuning Single-Phase Phase-Locked Loop Circuits

Dayane M. Lessa(ID), Michel P. Tcheou(ID), Cleiton M. Freitas$^{(\boxtimes)}$(ID), and Luís Fernando C. Monteiro(ID)

Rio de Janeiro State University, Rio de Janeiro, RJ 20550-900, Brazil
{cleiton.freitas,mtcheou,lmonteiro}@uerj.br

Abstract. Phase-Locked Loop (PLL) circuits have contributed to modernising electrical grids in different segments, such as distributed generation, identification and characterisation of phenomena related to power quality, localisation of faults, among others. This justifies the undergone research aiming at increasing their performance under certain conditions. Taking the example of the single-phase PLLs, researchers have worked out ways to cope with the characteristic doubly-frequency oscillation, which can undermine the performance of the frequency and phase tracking and compromise the extraction of the fundamental component from the input signal. In this sense, the present article aims at analysing a single-phase PLL circuit providing a methodology to adjust the control gains and minimise low-frequency oscillation. As for the analysed PLL, simulations in the time domain were carried out for modified versions of the well-known E-PLL, SOGI-PLL, and APF-PLL, all of them comprising notch filters and to cope with the doubly-frequency oscillation.

Keywords: Phase-Locked Loop · Orthogonal signal generators · Tuning methodology · Low-frequency oscillations

1 Introduction

Phase-Locked Loop (PLL) circuits are essential for instrumentation and control circuits that are required to be synchronised to the power grid. In fact, the use of PLL contributed to the modernisation of electrical networks in different segments, such as distributed generation, identification and characterisation of phenomena related to power quality, fault-finding on the power grid, among others [1].

When the PLL is applied in low-voltage or weak grids, the tracking of frequency and phase could be compromised by power quality issues, such as presence of unbalanced and harmonic components in the voltages and currents analysed [2,3]. These components may lead the PLL to present oscillating components in the tracked frequency, which could undermine the performance of

J. L. Afonso et al. (Eds.): SESC 2020, LNICST 375, pp. 132–151, 2021.
https://doi.org/10.1007/978-3-030-73585-2_9

PLL-dependent systems, e.g. grid-tied converters. Despite the fact that power quality plays an important role in the PLL performance, there are other issues which could not be neglected. For instance, in single-phase applications the general PLL presents a frequency oscillation at twice the frequency of the grid and it propagates inward, distorting internal signals, and compromising performance of the synchronisation circuit. An alternative to mitigate the problem is the use of an auxiliary signal in quadrature with the fundamental component of the input signal [4,5]. Then for an input signal composed only of the fundamental component, the single-phase PLL in steady-state has a similar behaviour to the three-phase PLL, eliminating the second-order harmonic component when the PLL is in a steady-state condition. Thus, such a harmonic component appears only when transients occur.

Another problem related to PLL is due to the difficulty in tuning the PLL controller gains. The analysis of non-linear models is still quite complex, and the problems related to them are still far from being solved, due to the complexity of the analysis of the mathematical model through second order differential equations [6]. Due to the analysis of transfer functions, the control theory used in most PLLs is strictly based on the linear control theory. In this case, a predefined model in the time domain is linearized for small-signals variations. One example of this approach is found in [6], where a method of analysing and simulating PLLs using linearisation is proposed.

The present work proposes an approach for tuning a single-phase PLL circuit, which is robust not only to variations in small signals, but also to variations in large signals. The methodology was explored in three different single-phase PLLs based on quadrature signal generators (QSG-PLLs) [7]: Enhanced Phase Locked Loop (E-PLL) [8–11], Second-Order Generalized Integrator (SOGI) PLL [12,13], and All-Pass filter PLL (APF-PLL) [14]. All the considered PLL structures were combined with a digital notch filter to mitigate second-order oscillations caused by distorted input signals [9,15].

Concerning the analysis presented in this article, it was considered a threshold of 100 ms, corresponding to 6 cycles of the fundamental component, for the settling time. In other words, any set of control parameters that lead to longer convergence time was disregarded. It was also disregarded the settings in which the output signals presented THD over 2%. It is important to highlight that there must be specific applications where these thresholds are prohibitive, but here it is used for simplicity.

2   Dynamic Models of the PLLs

This section presents the three different PLLs considered in the paper and also provides some discussion about their dynamic models. It is important to mention that, differently from the classical structures of the presented PLLs, this research considered the use of an adaptative notch filter to cope with the doubly-frequency components that may appear in the inner signals. The notch filter corresponds to a band-stop filter with a narrow band, that is, a high value for the quality factor.

The quality factor is the ratio between the resonant frequency and bandstop bandwidth. Thereby, the higher is the quality factor, the narrower is the band-stop range, and more accurately a specific frequency component is removed from the signal.

Figure 1 presents the frequency response of a second-order notch filter, whose transfer function is represented by (1), and the resonant frequency of the filter is represented by ω_n [16]. The quality factor was set as $Q = 1$, $Q = 5$, $Q = 10$ and $Q = 35$ and the component frequency to be attenuated was set 120 Hz. As can be seen, although the stopband is narrower for higher Q values, the attenuation at the resonant frequency increases with the reduction of the quality factor. Due to this factor, it was chosen to work with the quality factor $Q = 1$.

$$H_n = \left(\frac{s^2 + \omega_n^2}{s^2 + \frac{\omega_n}{Q}s + \omega_n^2} \right) \tag{1}$$

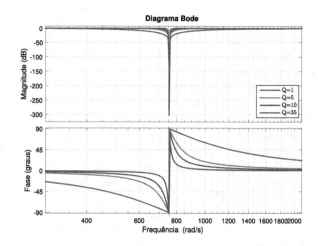

Fig. 1. Frequency response of the second-order notch filter

In order to attenuate the second harmonic component, a second order notch filter was designed and used in all PLLs models presented in this work. This filter attenuates second harmonic oscillations in the output signal, reducing the ripple in the PLL internal signals, and also improving the PLL convergence time.

For the mathematical description presented in this section form this point on, it is important to point out that it is neglected the possible effects caused by the notch filters. Such an assumption was considered once they were designed to filter out only the second harmonic component, without reflecting at the DC and fundamental components. Generally, the notch filter is used for attenuating a specific harmonic component present in the input signal and produces less phase lags.

2.1   E-PLL Dynamic Model

The Enhanced Phase Locked-Loop (E-PLL), proposed by Karimi [8–11], consists of two control loops. The first one aims to extract the frequency and phase-angle of the input signal, while the second loop obtains the signal amplitude. Figure 2 shows the E-PLL block diagram, including the notch filter in the control loop to track the phase-angle of the input signal.

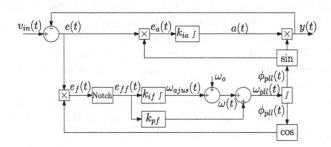

Fig. 2. E-PLL Diagram with a notch filter in the phase-loop

The amplitude control loop acts on the amplitude error, represented by $e_a(t)$, using an integral controller with gain k_{ia}. On the other hand, the phase-loop acts on the phase error, $e_f(t)$, using a PI controller with proportional and integral gains equal to k_{pf} and k_{if}. The parameter ω_o can be understood as an initial condition, being used to reduce the PLL convergence time. Finally, the E-PLL output signal, $y(t)$, represents the fundamental component (amplitude, phase and frequency) of the input signal $v_{in}(t)$. It is important to highlight that the mentioned gains influence the dynamic of both control loops, since there is no decoupling between them.

One must note that although both control loops run simultaneously, the control loop of the frequency and phase-angle reaches the steady-state condition before the amplitude control loop. This is due to the fact that the frequency and phase-angle control-loop is able to reach the steady-state condition regardless of the amplitude signal. This feature is exploited in sequence.

Assuming that ω_1 is the fundamental frequency of the input signal, and ϕ_{pll} is the PLL estimated phase, the input and output-signals, respectively, are given by:

$$v_{in}(t) = V_{in} \sin(\omega_1 t); \tag{2}$$

$$y(t) = a(t) \sin\left(\phi_{pll}(t)\right). \tag{3}$$

The phase-loop error signal, $e_f(t)$, corresponds to the difference between the E-PLL input and output-signals, when this difference is multiplied by the term $cos(\phi_{pll})$. Thus, $e_f(t)$ was decomposed into three components as follows:

$$e_f(t) = \frac{V_{in}}{2}\left[\sin\left(\omega_1 t - \phi_{pll}(t)\right) + \sin\left(\omega_1 t + \phi_{pll}(t)\right)\right] - \frac{a(t)}{2}\sin\left(2\phi_{pll}(t)\right). \quad (4)$$

Assuming that ω_1 and ϕ_{pll} are always positive, one may see that $e_f(t)$ presents an average component when $\omega_{pll} = \omega_1$, but $\phi_{pll}(t) \neq \omega_1 t$. Other aspect that must be considered corresponds to the fact that the notch filter attenuates only the second-harmonic component. Thus, even with oscillating components at $e_{ff}(t)$, it is possible to assume that an steady-state condition is achieved when the average components at $e_{ff}(t)$ and $e_f(t)$ decreases to zero, i.e., the input and output-signals are synchronised.

After the convergence process of the phase loop, ϕ_{pll} tends to $\omega_1 t$ and the term $\sin(\omega_1 t - \phi_{pll})$ in (4) tends to disappear. Also in this situation, the term $\sin(\omega_1 t + \phi_{pll})$ tends to $\sin(2\omega_1 t)$ and (4) becomes:

$$e_f(t) = \frac{1}{2}\left[V_{in} - a(t)\right]\sin(2\omega_1 t). \quad (5)$$

On the other hand, the amplitude control loop is dynamically modified while the average of signal $e_a(t)$ is different from zero. Expanding this control loop, $e_a(t)$ is given by:

$$e_a(t) = \frac{a(t)}{2}\left[1 - \cos\left(2\phi_{pll}(t)\right)\right]$$
$$- \frac{V_{in}}{2}\left[\cos\left(\omega_1 t - \phi_{pll}(t)\right) - \cos\left(\omega_1 t + \phi_{pll}(t)\right)\right]. \quad (6)$$

An average component appears more evident at $e_a(t)$ when ϕ_{pll} is equal to $\omega_1 t$. Indeed, when such condition is reached, $e_a(t)$ is reduced to:

$$e_a(t) = \frac{1}{2}\left[a(t) - V_{in}\right]; \quad (7)$$

which means that the amplitude control loop is not at the steady-state condition when ω_{pll} is equal to ω_1. Such condition is only reached when $a(t)$ becomes equal to V_{in}. Furthermore, back to the frequency and phase-angle control-loop, note that an oscillating component still remains at $e_f(t)$ while $a(t)$ is different from V_{in}, as shown in (5). Thus, such analysis reinforces the assumption of considering an overall steady-state condition to the E-PLL only after the amplitude control loop reaches such condition.

2.2   SOGI-PLL Dynamic Model

The Second Order Generalized Integrator (SOGI) structure integrated into the PLL circuit results in the synchronising circuit known as SOGI-PLL, as shown in Fig. 3. Observe that the SOGI-PLL corresponds to the conventional q-PLL [17], gathered with an auxiliary loop to generate the control signals $v_\alpha(t)$ and $v_\beta(t)$.

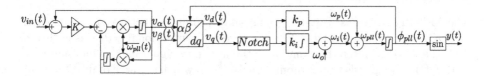

Fig. 3. SOGI-PLL Diagram with a notch filter in the phase-loop

The input signal is represented by $v_{in}(t)$ as indicated in (2), and with $v_\alpha(t)$ and $v_\beta(t)$ can be arranged as follows:

$$\frac{dv_\alpha(t)}{dt} + K\omega_{pll}(t)v_\alpha(t) = \omega_{pll}(t)\left[Kv_{in}(t) - v_\beta(t)\right];$$ (8)

$$\frac{dv_\beta(t)}{dt} = \omega_{pll}(t)v_\alpha(t).$$ (9)

In sequence, the control signal $v_q(t)$ is given by:

$$v_q(t) = v_\alpha(t)\cos\left(\phi(t)\right) - v_\beta(t)\sin\left(\phi(t)\right).$$ (10)

When the average component of $v_q(t)$ becomes equal to zero, the steady-state condition is reached. Thus, such condition occurs with $\omega_{pll}(t)$ synchronised with ω_1 and, moreover, the transient components of $v_\alpha(t)$ must be extinguished. Both dynamics occur simultaneously due to the feedback loop.

Expanding the SOGI loop and assuming that $\omega_{pll}(t)$ is synchronised with $\omega_1 t$, i.e., ω_{pll} is equal to ω_1, the relationship between $v_\alpha(t)$ and $v_{in}(t)$ is given by:

$$\frac{d^2v_\alpha(t)}{dt^2} + K\frac{dv_\alpha(t)}{dt} + \omega_1^2 v_\alpha(t) = K\omega_1 v_{in}(t).$$ (11)

Solving the second-order ordinary-equation, the steady-state solution of $v_\alpha(t)$ is $v_{in}(t) = V_{in}sin(\omega_1 t)$ with $v_\beta(t) = -V_{in}cos(\omega_1 t)$. The dynamics of $v_\alpha(t)$ and $v_\beta(t)$ as a function of the parameter K was described in [18], where $K \geq 2$ leads to an over-damped response. A detailed dynamic analysis to establish some constraints for tuning K was introduced by [13]. In this work, based on the aforementioned proposals combined with some preliminary results, it was considered a gain $K = 3$. Compared to E-PLL, SOGI-PLL has better immunity to distortions in the input signal, $v_{in}(t)$. This occurs due to integrators in SOGI structure acting as low-pass filters. Thus, the $v_\alpha(t)$ and $v_\beta(t)$ signals have less harmonic distortion compared to that observed in $v_{in}(t)$. On the other hand, such feature may lead to worse dynamics in comparison to the one obtained with E-PLL. Thus a commitment must be established for tuning its internal parameters as proposed in this work.

2.3   APF-PLL Dynamic Model

The APF-PLL (All-Pass Filter PLL) presents a similar structure when compared to SOGI-PLL, as illustrated in Fig. 4, where the conventional q-PLL is gathered with the All-Pass filter that generates the auxiliary signal $v_\beta(t)$. One may note that $v_\alpha(t)$ corresponds to the input signal, $v_{in}(t)$, which one was omitted. Essentially, the SOGI loop was replaced by an All-Pass Filter (APF) tuned at $\omega_{pll}(t)$. When the steady-state condition is reached, $\phi_{pll}(t)$ is synchronised with $\omega_1 t$, and $v_\beta(t)$ is in quadrature with $v_\alpha(t)$ as expected. Remind that ω_1 is the frequency of $v_{in}(t)$.

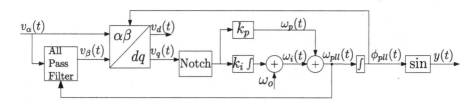

Fig. 4. Diagram blocks of APF-PLL including a nocth filter in the phase-loop

Figure 5 shows the All-Pass Filter diagram-blocks, where ω_{pll} is applied to determine $v_\beta(t)$ as a function of $v_\alpha(t)$, which leads to the following transfer function:

$$G_\beta(s) = \frac{v_\beta(s)}{v_\alpha(s)} = -\frac{s - \omega_{pll}}{s + \omega_{pll}}. \tag{12}$$

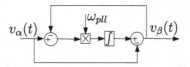

Fig. 5. Block diagrams of the All-Pass Filter

Expanding the block diagrams to obtain $v_\beta(s)$ as a function of $v_\alpha(s)$ and converting the expression of $v_\beta(s)$ to time-domain, $v_\beta(t)$ is given by:

$$v_\beta(t) = \frac{V_{in}}{\omega_1^2 + \omega_{pll}^2}\left[2\omega_1\omega_{pll}e^{-\omega_{pll}t} - \left(\omega_1^2 - \omega_{pll}^2\right)\sin(\omega_1 t) - 2\omega_1\omega_{pll}\cos(\omega_1 t)\right]. \quad (13)$$

When the average value of $\omega_{pll}(t)$ becomes equal to ω_1, and neglecting its ripple, the function of $v_\beta(t)$ can be reduced as follows:

$$v_\beta(t) = V_{in}\left[e^{-\omega_1 t} - cos(\omega_1 t)\right]. \quad (14)$$

At steady-state $v_\beta(t)$ becomes equal to $-V_{in}cos(\omega t)$, i.e., $v_\beta(t)$ becomes a sinusoidal waveform lagged $90°$ by $v_\alpha(t)$ as expected. In sequence, the APF-PLL presents a phase-loop similar as the one introduced in SOGI-PLL, where the control signal $v_q(t)$ is equal to:

$$v_q(t) = \sin\left(\phi_{pll}(t)\right)V_{in}\left[\cos(\omega_1 t) - e^{-\omega_1 t}\right] + V_{in}\sin(\omega_1 t)\cos\left(\phi_{pll}(t)\right). \quad (15)$$

At steady-state, with $\omega_{pll}(t)$ equals to ω_1 and the transient response of $v_\beta(t)$ extinguished, $v_q(t)$ is only comprised by an average component equal to zero. Thus, in comparison to the SOGI-PLL, the APF-PLL presents different dynamics to reach the steady-state condition. Based on the aforementioned conditions to be at steady-state, $\phi_{pll}(t)$ must be synchronised with $\omega_1 t$, as well as the transient response of $v_q(t)$ must be extinguished.

Based on the introduced PLL dynamics models, one may note that tuning the internal parameters of the PLLs through linear model approximations is unfeasible. Therefore, this paper presents an approach for tuning the internal parameters as described in sequence.

3   Proposed Approach for Tuning the Internal Parameters of the PLLs

In literature, one may see some proposals for tuning the internal parameters of the PLLs as, for instance, those introduced in [19–23]. Nonetheless, all of them were based on linear model approximations [24,25], which consists of predefined time-domain models valid for small signal variations. Thus, as highlighted in Sect. 2, due to the considerable assumptions and constraints, such models lead to limited-range solutions.

Therefore, it is proposed in this work an approach where the PLLs were submitted to a predefined set of parameters (internal gains). In sequence, data about the convergence speed of the internal loops and harmonic distortion of the output waveform were collected. The flowchart of Fig. 6 summarises the methodology for tuning the PLL control throughout this work [26]. The general idea was to conduct a series of tests considering different setting points (different values

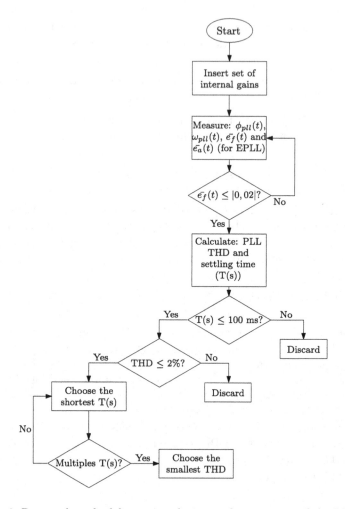

Fig. 6. Proposed method for tuning the internal parameters of the PLLs

for the control gains, k_{pf}, k_{if} and k_{ia}) for the PLLs defined in Figs. 2, 3 and 4. During these tests, the effects caused by the control gains in some performance parameters were analysed and used to indicate the best combinations to comply with some constraints (later explained). As can be seen, two performance parameters were considered during this process: the settling time T and the total harmonic distortion THD of the output signal y.

The settling time (convergence speed) was calculated as a function of the average component of the loop-errors. In the cases of the SOGI- and APF-PLLs, only the phase-loop error, $\overline{e_f}(t)$, was considered, whereas, for the E-PLL, it is also considered the average component of the amplitude-loop error, $\overline{e_a}(t)$. To compute the average components of these errors, it is considered a windowed moving average filter with a time span of (1/120) s, which corresponds to, approximately, 8.33 ms. Such assumption was done once $v_q(t)$ is comprehended by a dc-component plus even-order harmonics (2ω, 4ω, 6ω, ...), being ω the fundamental frequency. For determining the time instant where the PLL achieved steady-state condition, it was considered indicated in the flowchart by $|e_f(t)| \leq 0.02$. In other words, the settling time was counted from the beginning of the simulation until the phase error fades under 0.02. This threshold was empirically chosen by analysing the phase signal for different conditions. The THD of the output signal, on the other hand, was computed considering the PLLS in a steady-state condition. A boundary condition was established for reducing the number of results presented in the following sections. In this case, the control settings that led to settling time higher than 100ms and/or THD over 2% were simply discarded. Notice that, this boundary condition was randomly chosen, as this paper does no focus on any specific application. Nonetheless, depending on the application, these limits could be either relaxed or sharpened.

It is important to notice that, in a realistic scenario, where the voltage in a single-phase grid is usually distorted, particularly by the 3rd, 5th, and 7th harmonic components, the internal PLL signals necessarily present oscillating components even in a steady-state condition. These components tend to compromise the phase-tracking performance. For this reason, the tuning of the three PLLs was conducted considering a high-distorted input signal comprising fundamental 3rd, 5th, and 7th harmonic components.

3.1 E-PLL Tuning

In sequence, it is applied the proposed approach for tuning the internal gains of the E-PLL. One may note that E-PLL is composed of 2 internal loops (amplitude-loop and phase-loop), and, thus, it was extended the proposed method considering them both. Based on this, it has applied, firstly, to the phase-loop and, in sequence, to the amplitude-loop. The convergence speed of the E-PLL phase-loop is shown in Fig. 7. Basically, the set of gains were arranged with k_{pf} constant and a range of different values for k_{if} and k_{ia}, such that $0 < k_{if} < 2000$ and $20 < k_{ia} < 120$. One may note that the settling time is lower than 0.02 s for $k_{pf} = 34$ (Fig. 7(d)). Thus, it considered $k_{pf} = 34$ and it has done once more the proposed method looking to the settling time of the amplitude-loop. The acquired results are illustrated in Fig. 8. As expected, one may note a higher settling time to achieve the steady-state condition. Taking into consideration the k_{pf} previously tuned through the phase-loop analysis, note that the acceptable results (settling time lower than 0.05s) are limited in $0 < k_{if} < 950$ and $118 < k_{ia} < 120$ (see Fig. 8(d)). In sequence, the THD of the output signals were calculated and, based on the obtained results, the E-PLL gains were defined as $k_{pf} = 34$, $k_{if} = 800$, $k_{ia} = 120$. Through them, the settling time of the amplitude-loop was equal to 49.9 ms with an output-signal THD of 1.68%.

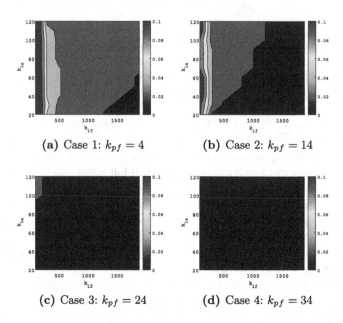

(a) Case 1: $k_{pf} = 4$ (b) Case 2: $k_{pf} = 14$

(c) Case 3: $k_{pf} = 24$ (d) Case 4: $k_{pf} = 34$

Fig. 7. Convergence speed of the E-PLL phase-loop

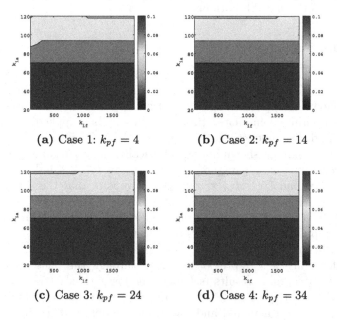

(a) Case 1: $k_{pf} = 4$ (b) Case 2: $k_{pf} = 14$

(c) Case 3: $k_{pf} = 24$ (d) Case 4: $k_{pf} = 34$

Fig. 8. Convergence speed of the E-PLL amplitude-loop

3.2 SOGI-PLL Tuning

For tuning the SOGI-PLL the procedure was done based on inspection of the
PLL phase-loop performance, with the constraint that the output-signal THD
must be lower than 2%. The set of the phase-loop gains was composed by a
finite range of values. Initially, the integral and proportional gains of the phase-
loop were limited to $5 < k_i < 2000$ and $5 < k_p < 2000$, with a step of 20
units. However, as shown in Fig. 9a, the contour region that delimits the shortest
settling time in the model proved to be very narrow and a second graph was
generated in order to obtain a better view of this band.

The new range of gains was set as $600 < k_i < 3000$ and $60 < k_p < 110$, with
a fixed step of 5 units for both gains. Then the circuit was inspected again in
order to generate a new and smaller database containing possible combinations
of gains in order to verify their influence on the control circuit.

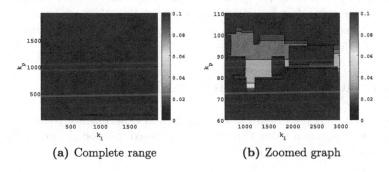

(a) Complete range   (b) Zoomed graph

Fig. 9. Convergence speed of the SOGI-PLL phase-loop

For a better observation, a new region presenting the convergence speed in
function of the phase-loop gains is shown in Fig. 9b, where it is possible to observe
that yellow area covers the best and shortest synchronisation times of the circuit.
Thus, in sequence, a new database was generated to identify those who presents
an output-signal waveform with THD lower than 2%. Through inspection, it
was found that the optimal gains for this particular analysis are referred to $k_i =$
1225 and $k_p = 75$. Through the selected gains, the SOGI-PLL with notch filter
presented an convergence speed of 66.4 ms and a THD of 0.68%.

3.3 APF-PLL Tuning

In comparison to the SOGI-PLL, the same procedure was applied to the APF-
PLL. Initially, informations of the settling time was collected with the set of
gains in a range defined by $5 < k_p < 1000$ and $5 < k_i < 1000$. Each set was
updated with a fixed step of 10 units. Based on the collected results, a contour
graphic illustrated in Fig. 10b was generated. One may note that all of the results
presented a settling time lower than 100 ms. Due to the lack of visibility of the

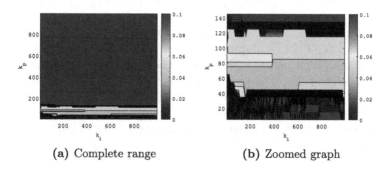

(a) Complete range   (b) Zoomed graph

Fig. 10. Convergence speed of the APF-PLL phase-loop

initial contour region and aiming to a better observation of the region containing the shortest PLL convergence speed, a second graph was generated with a range limited for the gains, $5 < k_i < 1000$ and $5 < k_p < 150$. Through inspection, the chosen gains correspond to $k_p = 45$ and $k_i = 425$, reaching an accommodation time of 49.9 ms and a THD of 0.62%.

4   Simulations and Experimental Results

In this section it is presented the simulation and experimental results of the 3 PLLs, with the chosen gains through the proposed method. In all of the test cases, the input signal was given by:

$$v_{in}(t) = sin(\omega_1 t) + \frac{1}{3}sin(3\omega_1 t) + \frac{1}{5}sin(5\omega_1 t) + \frac{1}{7}sin(7\omega_1 t), \qquad (16)$$

being ω_1 60 Hz fundamental frequency. Furthermore, it was only considered the transient when the PLLs were turned on.

The experimental results were obtained through a digital signal processor (DSP), with a internal clock of 150 MHz and sampling frequency of 40 kHz. The output signals were modulated through PWM (Pulse Width Modulation) technique. Small passive filters were used to attenuate the high-frequency component resulted from the switching frequency. On the other hand, the simulation results were obtained through Octave program.

4.1   Simulation and Experimental Results of the E-PLL

Figure 11a represents the phase-loop error before $(e_f(t))$ and after the notch filter $(e_{ff}(t))$. In addition, it is also illustrated the input and output signals. As expected, the notch filter did not compromised the phase-loop dynamics, reaching the steady-state condition in a time period lower than 2 cycles of the fundamental frequency. In sequence, Fig. 11b illustrates the amplitude-loop error $(e_a(t))$ together with the spectrum harmonic of the input and output signals.

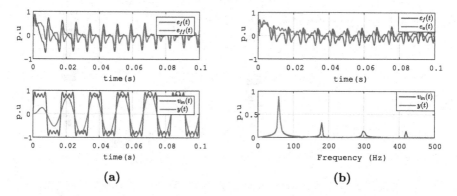

Fig. 11. E-PLL simulation results: (a) Phase-loop error with and without notch filter, including the input and output signals and (b) Amplitude-loop error with the harmonic spectrum of the input and output signals.

One may note that the amplitude-loop error reached its steady-state in a time period lower than 3 cycle periods.

Other aspect that should be highlighted is the notch filter performance, which one was designed to smooth an amount of the oscillating components in the phase-loop only. Thus, for keeping the THD of the output signal below 2%, both integrators contributed as low-pass filters, attenuating the oscillating components of $e_{ff}(t)$ and $e_a(t)$.

In sequence, the experimental results are illustrated in Fig. 12, evaluating the previous results obtained through simulations. As expected, the phase-loop, even with the notch filter, presents a faster dynamics in comparison to the amplitude-loop as shown in Fig. 12b. Moreover, from Fig. 12d, one may note that the output signal presents an small third harmonic component, with $e_a(t)$ having a stronger influence due to their oscillating components. Nevertheless, its THD is 2% below, which was previously defined.

4.2   Simulation and Experimental Results of the SOGI-PLL

In sequence, there are the simulation and experimental results of the SOGI-PLL. Initially, Fig. 13 presents the acquired results from simulation, where Fig. 13a shows the phase-loop error before ($v_q(t)$) and after ($v_{qf}(t)$) the notch filter with the input and output signals. The harmonic spectrum of the input and output signals are shown in Fig. 13b. One may note that, despite of $y(t)$ does not present the harmonic contents of $v_{in}(t)$, $y(t)$ still has an small 3rd harmonic due to the residual oscillating components in $v_{qf}(t)$.

Figure 14 introduces the experimental results at steady-state. From Fig. 14b note that, in comparison to $v_{in}(t)$, $v_\alpha(t)$ and $v_\beta(t)$ present a lower harmonic distortion due to the SOGI-loop integrators. Furthermore, the harmonic content in $v_\beta(t)$ is even lower in comparison to $v_\alpha(t)$. It occurs due to the double integrator action to generate $v_\beta(t)$. Nonetheless, both signals still present har-

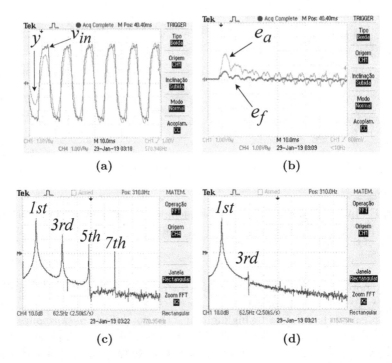

Fig. 12. E-PLL experimental results: (a) input and output signals, $v_{in}(t)$ and $y(t)$; (b) phase-loop error ($e_{ff}(t)$) and amplitude-loop error ($e_a(t)$)); (c) FFT of the input $v_{in}(t)$, and; (d) FFT of the output signal $y(t)$.

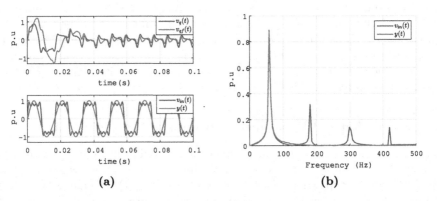

Fig. 13. SOGI-PLL simulation results: (a) Phase-loop error before ($v_q(t)$) and after ($v_{qf}(t)$) the notch filter; (b) Harmonic spectrum of the input and output signals.

monic distortion which reinforces the importance of the notch filter, combined with the integrator in the phase-loop, for reducing the oscillating-components propagation through the phase-loop for generating the output signal, $y(t)$.

In sequence, the harmonic spectrum of the input and output signals are shown in Fig. 14c and d, respectively. As aforementioned in this section, $y(t)$ still has an smaller 3rd harmonic component, due to the oscillating components in the signal that corresponds to the phase-loop error $(v_{qf}(t))$. Nonetheless, $y(t)$ presents THD below than 1%, which is acceptable based on the initial constraints for tuning the internal parameters.

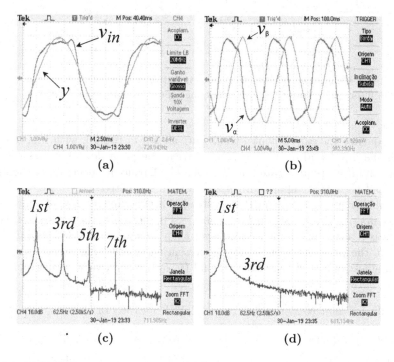

Fig. 14. SOGI-PLL experimental results: (a) input and output signals, $v_{in}(t)$ and $y(t)$; (b)quadrature auxiliary signals, $v_\alpha(t)$ and $v_\beta(t)$; (c) FFT of the input signal $v_{in}(t)$; and (d) FFT of the output signal $y(t)$.

4.3   Simulation and Experimental Results of the APF-PLL

Simulation results acquired from APF-PLL are shown in Fig. 15, where Fig. 15a presents the phase-loop error before $(v_{qf}(t))$ and after $(v_q(t))$ notch filter, with the input $(v_{in}(t))$ and output $(y(t))$ signals, whereas Fig. 15b illustrates the harmonic spectrum of $v_{in}(t)$ and $y(t)$. In comparison with SOGI-PLL, instead of having some differences for generating the auxiliary signals before the phase-loop, both PLLs presented a similar performance. It has occurred due to the proposed method for tuning the internal parameters of the loop-error, forcing both PLLs for presenting a similar performance to reach the steady-state condition with the output signal presenting THD below than 2%.

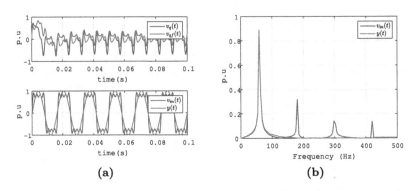

Fig. 15. APF simulation results: (a) Phase-loop error before $(v_q(t))$ and after $(v_{qf}(t))$ the notch filter; (b) Harmonic spectrum of the input and output signals.

In sequence the experimental results are shown in Fig. 16, with Fig. 16a presenting the input and output signals, whereas the phase-loop error before and after notch filter are illustrated in Fig. 16b. The harmonic spectrum of the input and output signals are shown in Fig. 16c and d, respectively.

It is important to comment that, instead of $y(t)$ having 3rd and 5th harmonic-components, its THD is still below than 1%. Furthermore, the steady-state condition was reached in a time period lower than 3 cycle periods (50 ms), and one may note the phase-loop error $(v_q(t))$ presenting the 2nd harmonic component as the highest one, and the filtered signal $(v_{qf}(t))$ with a very small ripple.

Table 1. Summary of settling times and THD for the PLL models analysed

PLL Model	EPLL			SOGI-PLL		APF-PLL	
Filter	Notch			Notch		Notch	
Parameter	T_{phase}	T_{amp}	THD(%)	T_{phase}	THD(%)	T_{phase}	THD(%)
Simulation	41.6 ms	49.9 ms	1.687	66.4 ms	0.6844	49.8 ms	0.6934
Experimental	30 ms	49.9 ms	1.687	62 ms	0.6844	50 ms	0.6934

Table 1 summarises all PLLs models dynamics, showing that all PLLs with notch filter in the phase-loop reach the steady-state close to the third wave cycle, but with different harmonic distortion. While E-PLL and APF-PLL had similar dynamic in the settling time, SOGI-PLL and APF-PLL had better results regarding THD reaching a value below 1%, half of the proposed value. The E-PLL obtained a THD of 1.69%.

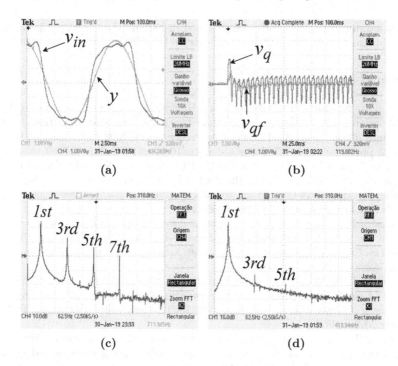

Fig. 16. APF-PLL experimental results: (a) input and output signals, $v_{in}(t)$ and $y(t)$; (b) phase-loop error before ($v_q(t)$) and after ($v_{qf}(t)$) notch filter; (c) FFT of the input signal $v_{in}(t)$; and (d) FFT of the output signal $y(t)$.

5   Conclusion

The present work proposes an approach for tuning single-phase PLLs to mitigate low-frequency oscillation. The proposed method was evaluated to three different PLLs, with all of them presenting orthogonal signal generators: E-PLL, SOGI-PLL, and APF-PLL.

Through simulation and experimental results one may note the capability of the proposed algorithm for tuning PLLs, even then presenting different dynamic behaviours. The employed notch filter was able to attenuate the oscillating component at 2ω without compromising the dynamics of the PLLs for converging to a steady-state condition in a time period specified in the proposed method. Nonetheless, in this work, the coefficients of the notch filter were constant, such that its capability for attenuating the specified frequency could be compromised if frequency deviations were considered. In future works it is intended to exploiting this issue, replacing the constant coefficients with variable ones, in function of the phase-loop output. Furthermore, other transient events must be considered for evaluating the proposed method.

Acknowledgement. This study was financed in part by the Coordenação de Aperfeiçoamento de Pessoal de Nível Superior – Brasil (CAPES) – Finance Code 001

References

1. Sakamoto, S., Izumi, T., Yokoyama, T., Haneyoshi, T.: A new method for digital PLL control using estimated quadrature two phase frequency detection. In: Proceedings of the Power Conversion Conference-Osaka 2002 (Cat. No. 02TH8579). vol. 2, pp. 671–676 (2002). https://doi.org/10.1109/PCC.2002.997599
2. Arruda, L.N., Silva, S.M., Filho, B.J.C.: PLL structures for utility connected systems. In: Conference Record of the 2001 IEEE Industry Applications Conference. 36th IAS Annual Meeting (Cat. No. 01CH37248), vol. 4, pp. 2655–2660 (2001). https://doi.org/10.1109/IAS.2001.955993
3. Modesto, R.A., da Silva, S.A.O.: Controladores srf aplicados em condicionadores ativos de potência em sistemas trifásicos a quatro-fios. UNOPAR Científica Ciências Exatas e Tecnológicas, **8**(1) (2009). https://revista.pgsskroton.com/index.php/exatas/article/view/609
4. Sepahvand, H., Saniei, M., Mortazavi, S.S., Golestan, S.: Performance improvement of single-phase PLLs under adverse grid conditions: an FIR filtering-based approach. Electric Power Syst. Res. **190**, 106829 (2021). https://doi.org/10.1016/j.epsr.2020.106829
5. Silva, S.M., Lopes, B.M., Filho, B.J.C., Campana, R.P., Bosventura, W.C.: Performance evaluation of PLL algorithms for single-phase grid-connected systems. In: Conference Record of the 2004 IEEE Industry Applications Conference, 2004. 39th IAS Annual Meeting, vol. 4, pp. 2259–2263 (2004). https://doi.org/10.1109/IAS.2004.1348790
6. Abramovitch, D.: Phase-locked loops: a control centric tutorial. In: Proceedings of the 2002 American Control Conference (IEEE Cat. No. CH37301). vol. 1, pp. 1–15 (2002). https://doi.org/10.1109/ACC.2002.1024769
7. Golestan, S., Guerrero, J.M., Vasquez, J.C.: Single-phase PLLs: a review of recent advances. IEEE Trans. Power Electron. **32**(12), 9013–9030 (2017). https://doi.org/10.1109/TPEL.2017.2653861
8. Karimi-Ghartema, M.: Enhanced Phase-Locked Loop Structures for Power and Energy Applications, chap. 1–5. Wiley-IEEE Press (2014). https://doi.org/10.1002/9781118795187
9. Karimi-Ghartemani, M.: Linear and pseudolinear enhanced phased-locked loop (EPLL) structures. IEEE Trans. Ind. Electron. **61**(3), 1464–1474 (2014). https://doi.org/10.1109/TIE.2013.2261035
10. Karimi-Ghartemani, M., Khajehoddin, S.A., Jain, P.K., Bakhshai, A., Mojiri, M.: Addressing DC component in PLL and notch filter algorithms. IEEE Trans. Power Electron. **27**(1), 78–86 (2012). https://doi.org/10.1109/TPEL.2011.2158238
11. Karimi-Ghartemani, M., Khajehoddin, S.A., Jain, P.K., Bakhshai, A.: Derivation and design of in-loop filters in phase-locked loop systems. IEEE Trans. Instrum. Meas. **61**(4), 930–940 (2012). https://doi.org/10.1109/TIM.2011.2172999
12. Golestan, S., Monfared, M., Freijedo, F.D., Guerrero, J.M.: Dynamics assessment of advanced single-phase PLL structures. IEEE Trans. Ind. Electron. **60**(6), 2167–2177 (2013). https://doi.org/10.1109/TIE.2012.2193863
13. Xiao, F., Dong, L., Li, L., Liao, X.: A frequency-fixed SOGI-based PLL for single-phase grid-connected converters. IEEE Trans. Power Electron. **32**(3), 1713–1719 (2017). https://doi.org/10.1109/TPEL.2016.2606623

14. Thacker, T., Wang, R., Dong, D., Burgos, R., Wang, F., Boroyevich, D.: Phase-locked loops using state variable feedback for single-phase converter systems. In: 2009 Twenty-Fourth Annual IEEE Applied Power Electronics Conference and Exposition, pp. 864–870 (2009). https://doi.org/10.1109/APEC.2009.4802763

15. Hogan, D.J., Gonzalez-Espin, F.J., Hayes, J.G., Lightbody, G., Foley, R.: An adaptive digital-control scheme for improved active power filtering under distorted grid conditions. IEEE Trans. Ind. Electron. **65**(2), 988–999 (2018). https://doi.org/10.1109/TIE.2017.2726992

16. Sedra, A., Sedra, D., Smith, K., Smith, P.: Microelectronic circuits. Oxford Series in Electrical and Computer Engineering, pp. 678–696. Oxford University Press (1998). https://books.google.com.br/books?id=RcodQm6LaVEC

17. Rolim, L.G.B., da Costa, D.R., Aredes, M.: Analysis and software implementation of a robust synchronizing PLL circuit based on the pq theory. IEEE Trans. Ind. Electron. **53**(6), 1919–1926 (2006). https://doi.org/10.1109/TIE.2006.885483

18. Kulkarni, A., John, V.: A novel design method for SOGI-PLL for minimum settling time and low unit vector distortion. In: IECON 2013–39th Annual Conference of the IEEE Industrial Electronics Society, pp. 274–279 (2013). https://doi.org/10.1109/IECON.2013.6699148

19. Gardner, F.M.: Phaselock Techniques, 3rd edn, pp. 8–63. Wiley, Hoboken (2005). https://doi.org/10.1002/0471732699

20. Kroupa, V.F.: Phase Lock Loops and Frequency Synthesis, chap. 4. Wiley, Hoboken (2003). https://doi.org/10.1002/0470014105

21. Leonov, G., Kuznetsov, N., Seledzhi, S.: Nonlinear analysis and design of phase-locked loops. In: Rodić, A.D. (ed.) Automation and Control, chap. 7. IntechOpen, Rijeka (2009). https://doi.org/10.5772/7900

22. Lindsey, W.C., Simon, M.K.: Telecommunication Systems Engineering, chap. 2–3. Dover Books on Electrical Engineering, Dover Publications (1991). https://books.google.com.br/books?id=m2-tYwrNMPQC

23. Viterbi, A.J.: Principles of Coherent Communication, chap. 2–5. McGraw-Hill, New York (1966)

24. Encinas, J.: Phase Locked Loops, chap. 2-3. Microwave and RF Techniques and Applications. Springer, New York (2012)

25. Meyr, H., Popken, L., Mueller, H.: Synchronization failures in a chain of PLL synchronizers. IEEE Trans. Commun. **34**(5), 436–445 (1986). https://doi.org/10.1109/TCOM.1986.1096569

26. Lessa, D.M.: Uso combinado de filtros digitais com circuitos de sincronismo monofásico. Master's thesis, Universidade do Estado do Rio de Janeiro (2019). https://www.pel.uerj.br/bancodissertacoes/Dissertacao_Dayane_Lessa.pdf

Optimized Power System Voltage Measurements Considering Power System Harmonic Effects

Chukwuemeka Obikwelu$^{(\boxtimes)}$ and Sakis Meliopoulos

Electrical and Computer Engineering, Georgia Institute of Technology, Atlanta, GA 30332, USA
{cobikwelu3,sakis.m}@gatech.edu

Abstract. The paper presents a real-time error correction method that accurately estimates the primary quantity of an inductive instrument transformer – that is, a voltage or current transformer – based on a Dynamic-State-Estimation (DSE) optimization approach.

The DSE error-correction method uses a high-fidelity model of the voltage transformer (VT) instrumentation channel (IC) and the Unconstrained-Weighted-Least-Squares (UWLS) optimization approach to accurately estimate the VT primary voltage; thus, correcting for errors generated in the instrumentation channel. Specifically, the method involves obtaining discretized measurements of the VT load voltage through a continuous monitoring process. The VT primary voltage is then continuously estimated by the method based on fitting the continuously obtained measurement samples to the high-fidelity VT model. The continuously estimated primary voltage can then be made readily available for metering, protection, and control functions. The method can be implemented using microprocessor technology, for example, in Merging Unit (MU) applications for digital power substation architectures.

The paper will focus on evaluating the performance of the DSE-based error-correction method, applied to inductive voltage transformer instrumentation channel (VTIC) applications, in the estimation of primary voltage signals distorted by harmonics. The evaluation process will employ multiple Fourier analysis techniques.

Keywords: VT · DFT · DSE · Optimization

1 Introduction

Harmonics in power systems are generally caused by nonlinear loading and device operations that draw non-sinusoidal currents from sinusoidal voltage sources – resulting in the injection of parasitic harmonic power into the power system, causing power quality problems. Harmonics circulation occurs primarily in distribution power systems. This is due to the nonlinear loads that are directly interconnected to distribution systems – e.g., industrial loads, and distributed generation (DG). Furthermore, with increasing DG connections, the number of inverter-based technologies is also increasing. The power electronics associated with these technologies can generate various harmonics

J. L. Afonso et al. (Eds.): SESC 2020, LNICST 375, pp. 152–171, 2021.
https://doi.org/10.1007/978-3-030-73585-2_10

in distribution systems. Other sources of harmonics include electric machine energization events, and operating transformers or electric machines in the nonlinear regions of their excitation-characteristics during stressed power system conditions. While transient events, such as equipment energization, can introduce harmonics onto the power system, harmonic effects are generally classified as steady-state phenomena. It is thus important to monitor power system harmonic levels as a prerequisite for and in support of implementing power quality improvement strategies. The VT is indispensably essential to this monitoring process.

Voltage transformers (VTs) provide scaled down replicas of the system-level voltage to low-voltage instrumentation devices. For example, typical VT secondary voltages used in protective relaying are 66.4 V, 115 V, and 120 V. The "instrumentation channel" delivers the scaled replica voltage to the connected devices, where it can then be used for metering, protection, and/or control schemes. The conventional or inductive voltage transformer instrumentation channel (VTIC) comprises the instrument transformer core, primary and secondary windings, control cables, electric loads, filters, and A/D converters. It is essential that the VTIC delivers accurate representations of the power system voltage to the connected devices during normal and disturbed system operations, otherwise the associated protection, control and/or monitoring functions may be compromised – posing risks to the power system.

Except for the scaling down, the replica or reproduced voltage should ideally resemble the primary voltage – in terms of waveshape, frequency, phase-shift, and harmonic content. However, instrumentation channel limitations and dynamics may cause the reproduced voltage to deviate from the ideal, and thus not represent a perfect replica of the primary voltage.

It is important to note that the transfer function of the VT behaves like a low-pass filter, causing higher harmonic signal components to experience varying degrees of attenuation. Furthermore, when harmonic content frequencies match any of the natural frequencies of the instrumentation channel, resonance effects (characterized by harmonic voltage swells) may significantly distort the secondary voltage. The resonance effects are caused by interactions within the VTIC, between the leakage inductances, magnetizing branch inductance and parasitic capacitances, in response to the harmonic content of the driving input or primary voltage. Therefore, the attenuation or swelling of harmonic voltage signal components may cause secondary voltage distortions, such that the secondary voltage no longer accurately represents a scaled-down replica of the primary voltage. More specifically, the harmonic content of the secondary voltage may no longer adequately represent the harmonic content of the primary voltage due to instrumentation channel limitations – thus, signifying the generation of channel errors. The inability of the VT to accurately reproduce the harmonic content present in primary voltage and thus present in the power system can be problematic for power quality monitoring.

The earlier mentioned instrumentation channel dynamics, which can contribute to the generation of error effects within the channel, generally refer to interactions between power system factors and instrumentation channel factors. These interactions are exacerbated by the limitations of the instrumentation channel, such as magnetic core size, etc. Examples of related power system factors are high source X/R ratio, system topography, and harmonics; examples of related VTIC factors are magnetic core characteristics, burden level, parasitic capacitances, leakage inductances and ferro-resonance suppression circuit parameters. The generated channel error effects include: magnetic core saturation; harmonic distortions in the secondary signal; transient secondary waveform irregularities, as occurs in Coupled-Capacitor-VT (CCVT) applications – where the discharging of CCVT internal energies, after fault-related system voltage collapse, results in nontrivial transient secondary waveform distortions.

It is therefore valuable to develop online methods that reliably correct for VTIC errors, which cause inaccuracies in both the fundamental and harmonic components of the secondary voltage measurement. The method studied in this paper is intended to achieve this online correction and provide optimized power system voltage measurements for connected secondary subsystem functions. Earlier related work is covered in [4] and the application of the method for current transformer instrumentation channel (CTIC) error correction is covered in [5] and [12].

2 Literature Review

Historically, different methods have been applied to correct for instrument transformer instrumentation channel (ITIC) error generation, although many of these methods focus on error-correction for current transformers (CTs). The method in [1] provides compensation for CCVTIC transient errors based on the digital inversion of the CCVT transfer function. This approach does not use a detailed CCVT model, is highly vulnerable to resonance effects, and neglects saturation possibilities. [2] proposes an output tracking VTIC error-correction method, based on repetitive learning control, for reproducing the primary voltage. The performance of the method may be jeopardized by serious saturation effects and the poor selection of controller gains. [3] proposes a CCVT error-correction method that estimates the primary voltage based on nonlinear least-squares fitting. The method may be vulnerable to convergence problems (from poor state vector initialization), and to errors from saturation effects and system harmonics unaccounted for in its fitting model. None of the reviewed methods were applied in cases where the primary voltage (to be reconstructed or estimated) is corrupted with harmonics – thus, how they would perform in such cases is uncertain. Furthermore, the reviewed methods may be suitable for control functions but not for protective relaying since their respective algorithms are not structured for real-time implementation.

This paper presents a new method for estimating the VT primary voltage on a sample-by-sample basis, using continuously obtained sample measurements of the VT load voltage. The method involves modeling the VTIC in detail and applying the unconstrained weighted least-squares (UWLS) dynamic state estimation (DSE) method to accurately estimate the primary voltage. The method is designed for real-time implementation, performs robustly with noisy inputs, and is immune to remnant flux effects. The paper will focus on evaluating the performance of the method in cases where the primary voltage is corrupted with harmonic components.

3 Methodology

The methodology involves a series of model development steps leading up to the formulation of the estimation problem. Different Least-Squares methods can be applied to solve the problem, but the presented methodology uses the Unconstrained Weighted Least-Squares (UWLS) technique.

3.1 VT Instrumentation Channel Electrical Circuit Model

The device or system under study, that is, the instrumentation channel, is first represented by a detailed electrical circuit. The circuit is parametrized to account for the unknown state variables. The device is then mathematically modeled by a system of algebraic, nonlinear, and differential equations, expressed in terms of the state variables. The system of equations is generally derived based on: applying Kirchhoff's Current and Voltage Laws at the nodes and within the loops of the electrical circuit model, respectively; applying Faraday's Law of Induction to relate the internal voltage with the flux; and, applying a high-order equation to model the nonlinear relationship between the magnetizing current and the flux.

Figure 1 shows the non-ideal or detailed high-frequency VTIC circuit model, including parasitic capacitances (C_1, C_2, C_3 and C_4). For the study, the following simplifications are made to the provided electrical circuit model:

1. The magnetizing branch current becomes insignificant in the high-frequency range. Coupling the latter point with the assumption that the VT is operating within the linear region of its excitation curve, the magnetizing branch may be neglected without consequences.
2. In the high-frequency range, it is necessary to model the transformer leakage reactances. But the parasitic capacitances may be neglected unless their effects have been found to be significant. Thus, there are generally two high-frequency range models, both including the transformer leakage reactances – one with parasitic capacitances and one without them. It may be important to model the parasitic capacitances depending on the harmonic effects being considered. For this study, the resonance effects of the parasitic capacitances are negligible for the range of harmonics considered – therefore, the capacitances are neglected.

The resulting circuit model is thus a high-frequency model, neglecting parasitic capacitance effects. The system of nonlinear, algebraic, and differential equations formed from the circuit model in Fig. 1 comprises 19 equations and 19 unknowns or state variables [4]. The above simplifications to the circuit model result in a system of algebraic and differential equations, comprising 11 equations and 11 state variables. The state vector is shown in (1), representing the minimum number of states that can be applied to fully describe the circuit model.

$$\mathbf{x}_1 = [v_1(t)\ \ v_2(t)\ \ v_3(t)\ \ v_4(t)\ \ v_5(t)\ \ v_6(t)\ \ e_C(t)\ \ i_{L1}(t)\ \ i_{L2}(t)\ \ i_{L3}(t)\ \ i_{L4}(t)\ ]^T \quad (1)$$

The voltages and currents in (1) are shown in Fig. 1. Note that the intended connection for the VT in this analysis is between phase and ground, and thus node 2 in Fig. 1

is grounded. Figure 2 presents the system of algebraic and differential equations that mathematically describes the VTIC circuit model shown in Fig. 1, after simplification. The stabilizing conductances (i.e., g_{s1}, g_{s2}, g_{s3} and g_{s4}), which are connected in-parallel with the circuit inductances, are not included in Fig. 1 for simplification purposes – but they are accounted for in the system of equations shown in Fig. 2. Furthermore, g_B represents the load conductance, and M_{34} represents the mutual inductance of the two-way secondary cabling.

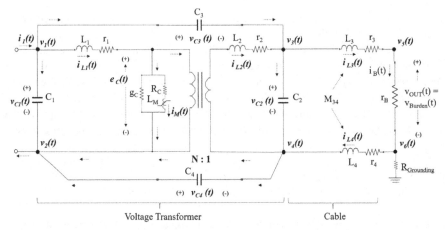

Fig. 1. High-frequency, non-ideal VTIC electrical circuit model, featuring parasitic capacitances.

The quadratization of the circuit model, as performed in related work in [4] and [5], is not needed in this analysis due to the applied simplifications. More specifically, the reasonable assumption of linear device operation and neglecting the magnetizing branch due to analysis in the high-frequency range remove the equations for modeling nonlinear excitation behavior from the mathematical model. Therefore, the next step is to use the mathematical model provided in Fig. 2 to develop the dynamic measurement models.

3.2   VT Instrumentation Channel Measurement Models

Dynamic Measurement Model (DMM)
The DMM framework mathematically associates the obtained measurements with the theoretical system model, plus error, as shown in (2).

$$\mathbf{z}(t) = \mathbf{h}(\mathbf{x}(t)) + \mathbf{\eta} \qquad (2)$$

1. $0 = -\{v_1(t) - v_2(t)\} + r_1 \cdot \left\{ i_{L1}(t) + g_{s1} L_1 \dfrac{di_{L1}(t)}{dt} \right\} + L_1 \dfrac{di_{L1}(t)}{dt} + e_c(t)$

2. $0 = -\dfrac{e_c(t)}{N} + L_2 \dfrac{di_{L2}(t)}{dt} + r_2 \cdot \left\{ i_{L2}(t) + g_{s2} L_2 \dfrac{di_{L2}(t)}{dt} \right\} + \{v_3(t) - v_4(t)\}$

3. $0 = -\{v_6(t) - v_4(t)\} + r_4 \cdot \left\{ i_{L4}(t) + g_{s4}\{L_4 \dfrac{di_{L4}(t)}{dt} - M_{34} \dfrac{di_{L3}(t)}{dt}\} \right\} + \left\{ L_4 \dfrac{di_{L4}(t)}{dt} - M_{34} \dfrac{di_{L3}(t)}{dt} \right\}$

4. $0 = -\{v_3(t) - v_5(t)\} + r_3 \cdot \left\{ i_{L3}(t) + g_{s3}\{L_3 \dfrac{di_{L3}(t)}{dt} - M_{34} \dfrac{di_{L4}(t)}{dt}\} \right\} + \left\{ L_3 \dfrac{di_{L3}(t)}{dt} - M_{34} \dfrac{di_{L4}(t)}{dt} \right\}$

5. $v_{out}(t) = v_5(t) - v_6(t)$

6. $0 = N \cdot \left\{ i_{L1}(t) + g_{s1} L_1 \dfrac{di_{L1}(t)}{dt} \right\} - \left\{ i_{L2}(t) + g_{s2} L_2 \dfrac{di_{L2}(t)}{dt} \right\}$

7. $0 = i_{L2}(t) + g_{s2} L_2 \dfrac{di_{L2}(t)}{dt} - \left\{ i_{L3}(t) + g_{s3}\{L_3 \dfrac{di_{L3}(t)}{dt} - M_{34} \dfrac{di_{L4}(t)}{dt}\} \right\}$

8. $0 = i_{L3}(t) + g_{s3} \cdot \left\{ L_3 \dfrac{di_{L3}(t)}{dt} - M_{34} \dfrac{di_{L4}(t)}{dt} \right\} - g_B \{v_5(t) - v_6(t)\}$

9. $0 = -\left\{ i_{L4}(t) + g_{s4} \cdot \left\{ L_4 \dfrac{di_{L4}(t)}{dt} - M_{34} \dfrac{di_{L3}(t)}{dt} \right\} \right\} + g_B \{v_5(t) - v_6(t)\}$

10. $0 = -\left\{ i_{L2}(t) + g_{s2} L_2 \dfrac{di_{L2}(t)}{dt} \right\} + \left\{ i_{L4}(t) + g_{s4}\{L_4 \dfrac{di_{L4}(t)}{dt} - M_{34} \dfrac{di_{L3}(t)}{dt}\} \right\}$

11. $0 = N \cdot \left\{ i_{L1}(t) + g_{s1} L_1 \dfrac{di_{L1}(t)}{dt} \right\} - g_B \{v_5(t) - v_6(t)\}$

Fig. 2. Mathematical model for the VTIC electric circuit model, after applying the proposed simplifications to the circuit in Fig. 1.

where $\mathbf{z}(t)$ is the measurement vector, comprising four different kinds of measurements to be discussed subsequently; $\mathbf{x}(t)$ is the state vector in (2); $\mathbf{h}(\mathbf{x}(t))$ is the overdetermined, theoretical system model, written in matrix equation form in (3); and, $\mathbf{\eta}$ is the error vector.

$$\mathbf{h}(\mathbf{x}(t)) = \mathbf{Y}_{eq} \cdot \mathbf{x}(t) + \mathbf{B}_{eq} \cdot \frac{d}{dt}\{\mathbf{x}(t)\} + \underbrace{\begin{bmatrix} \mathbf{x}(t)^T \cdot \mathbf{F}_{eq1(S \times S)} \cdot \mathbf{x}(t) \\ \vdots \\ \mathbf{x}(t)^T \cdot \mathbf{F}_{eqN_m(S \times S)} \cdot \mathbf{x}(t) \end{bmatrix}}_{\text{Quadratic Terms}} + \mathbf{C}_{eq}$$

(3)

where $\mathbf{Y}_{eq}$, $\mathbf{B}_{eq}$, $\mathbf{F}_{eq(1\text{-}Nm)}$ and $\mathbf{C}_{eq}$ are the coefficient matrices of the linear, differential, *quadratic* and constant parts, respectively; S is the number of states after quadratization; N_m is the number of elements in the measurement vector. Quadratic terms, and additional state variables, typically result from the quadratization process. But due to the applied simplifications, the measurement model does not have any quadratic terms; thus, the matrix formulation accounting for them in (3) above can be neglected without consequence. The overdetermined system model used in the DMM framework (that is, the DMM system model) is shown in Fig. 3.

1. $0 = -\{v_1(t) - v_2(t)\} + r_1 \cdot \left\{ i_{L1}(t) + g_{s1} L_1 \dfrac{di_{L1}(t)}{dt} \right\} + L_1 \dfrac{di_{L1}(t)}{dt} + e_c(t)$

2. $0 = -\dfrac{e_c(t)}{N} + L_2 \dfrac{di_{L2}(t)}{dt} + r_2 \cdot \left\{ i_{L2}(t) + g_{s2} L_2 \dfrac{di_{L2}(t)}{dt} \right\} + \{v_3(t) - v_4(t)\}$

3. $0 = -\{v_6(t) - v_4(t)\} + r_4 \cdot \left\{ i_{L4}(t) + g_{s4}\{L_4 \dfrac{di_{L4}(t)}{dt} - M_{34} \dfrac{di_{L3}(t)}{dt}\} \right\} + \left\{ L_4 \dfrac{di_{L4}(t)}{dt} - M_{34} \dfrac{di_{L3}(t)}{dt} \right\}$

4. $0 = -\{v_3(t) - v_5(t)\} + r_3 \cdot \left\{ i_{L3}(t) + g_{s3}\{L_3 \dfrac{di_{L3}(t)}{dt} - M_{34} \dfrac{di_{L4}(t)}{dt}\} \right\} + \left\{ L_3 \dfrac{di_{L3}(t)}{dt} - M_{34} \dfrac{di_{L4}(t)}{dt} \right\}$

5. $v_{out}(t) = v_5(t) - v_6(t)$

6. $0 = N \cdot \left\{ i_{L1}(t) + g_{s1} L_1 \dfrac{di_{L1}(t)}{dt} \right\} - \left\{ i_{L2}(t) + g_{s2} L_2 \dfrac{di_{L2}(t)}{dt} \right\}$

7. $0 = i_{L2}(t) + g_{s2} L_2 \dfrac{di_{L2}(t)}{dt} - \left\{ i_{L3}(t) + g_{s3}\{L_3 \dfrac{di_{L3}(t)}{dt} - M_{34} \dfrac{di_{L4}(t)}{dt}\} \right\}$

8. $0 = i_{L3}(t) + g_{s3} \cdot \left\{ L_3 \dfrac{di_{L3}(t)}{dt} - M_{34} \dfrac{di_{L4}(t)}{dt} \right\} - g_B\{v_5(t) - v_6(t)\}$

9. $0 = -\left\{ i_{L4}(t) + g_{s4} \cdot \left\{ L_4 \dfrac{di_{L4}(t)}{dt} - M_{34} \dfrac{di_{L3}(t)}{dt} \right\} \right\} + g_B\{v_5(t) - v_6(t)\}$

10. $0 = -\left\{ i_{L2}(t) + g_{s2} L_2 \dfrac{di_{L2}(t)}{dt} \right\} + \left\{ i_{L4}(t) + g_{s4}\{L_4 \dfrac{di_{L4}(t)}{dt} - M_{34} \dfrac{di_{L3}(t)}{dt}\} \right\}$

11. $0 = N \cdot \left\{ i_{L1}(t) + g_{s1} L_1 \dfrac{di_{L1}(t)}{dt} \right\} - g_B\{v_5(t) - v_6(t)\}$

12. $i_B(t) = g_B \cdot \{v_5(t) - v_6(t)\}$

13. $i_B(t) = i_{L2}(t) + g_{s2} L_2 \dfrac{di_{L2}(t)}{dt}$

14. $i_B(t) = i_{L3}(t) + g_{s3} \cdot \left\{ L_3 \dfrac{di_{L3}(t)}{dt} - M_{34} \dfrac{di_{L4}(t)}{dt} \right\}$

15. $i_B(t) = i_{L4}(t) + g_{s4} \cdot \left\{ L_4 \dfrac{di_{L4}(t)}{dt} - M_{34} \dfrac{di_{L3}(t)}{dt} \right\}$

16. $0 = v_2(t)$

17. $0 = v_6(t)$

Fig. 3. VTIC dynamic measurement model (DMM) system model.

The measurement vector $\mathbf{z}(t)$ comprises four kinds of measurements, namely:

- Actual measurements: The only actual measurement in the DMM is the output or load voltage. Equation (5) in Fig. 3 represents the *actual* measurement.
- Virtual measurements: This serves to represent the implementation a physical law, such as Faraday's Induction Law, Kirchhoff's Current Law (KCL) or Kirchhoff's Voltage Law (KVL). Equations (1) through (4) and (6) through (11) in Fig. 3 represent *virtual* measurements.
- Derived measurements: This is a measurement that is derived from an actual measurement. The derived measurement shares the same error as the actual measurement used for its derivation. Equations (12) through (15) in Fig. 3 represent *derived* measurements.
- Pseudo measurements: This is a hypothetical measurement, where its expected value can be guessed although with high uncertainty. Equations (16) and (17) in Fig. 3 represent *pseudo* measurements.

Algebraic Companion Form (ACF) Measurement Model

The ACF measurement model results from integrating the DMM using the Quadratic Integration method [6], which is in the family of Lobatto-Runge-Kutta numerical integration methods. The quadratic integration method does two key things:

- The system model equations are integrated assuming that the system states vary quadratically within each time step.
- Reformulates the nonlinear system model equations as a fully equivalent system of quadratic algebraic equations (by introducing additional state variables).

The integration time-step h, also the DSE time-step, is set to twice the sampling period (Ts) of the streaming measurements. The DMM is then integrated from time $t - h$ to t_m (that is, $t - h/2$) and from time $t - h$ to t. Thus, the measurement vector $\mathbf{z}(t,tm)$ now contains two sets of measurements: the first set of measurements at time t, stacked on-top of a second set of the same measurements but at time t_m (or one sampling period prior). The system model $\mathbf{h}(\mathbf{x}(t,t_m))$ is now expressed as a function of an expanded state vector, $\mathbf{x}(t,t_m)$. The state vector $\mathbf{x}(t,t_m)$ now has two sets of state variables: the first set of states at time t, stacked on-top of the second set of states at time t_m (or one sampling period prior).

The ACF measurement model is the final state of the model development process. The next step is to apply this measurement model in the formulation and resolution of the estimation problem. Given the ACF measurement model setup, the error-correction algorithm is setup such that, at each DSE timestep, it provides the best-estimates least-squares state vector solution at the current time t and at time $t - h/2 = t - Ts$. This setup provides the basis for the specification of a criterion for real-time execution, which the proposed method adheres to. The criterion is stated as follows: for real-time execution, the algorithm should provide the estimates at times t and $t - h/2$ (i.e., the present time and one sampling period prior) before the arrival of the next sample, which is one DSE time-step (that is, two sampling periods) away.

3.3 Problem Formulation and Solution Method

Using the ACF measurement model, the estimation problem is formulated as an unconstrained weighted least squares (UWLS) optimization problem – as shown in (4) and (5).

$$\min \ J(x(t, t_m)) = \sum_{j=1}^{N_{mm}} \left(\frac{z^{(j)}(t, t_m) - h^{(j)}(x(t, t_m))}{\sigma^{(n)}} \right)^2 = \sum_{j=1}^{N_{mm}} \left(\frac{\eta^{(j)}}{\sigma^{(j)}} \right)^2 = \sum_{j=1}^{N_{mm}} \left(s^{(j)} \right)^2$$

(4)

$$\min \ J = \mathbf{\eta}^T \cdot \mathbf{W} \cdot \mathbf{\eta}$$

(5)

where, $s^{(j)}$ represents the normalized residual for the jth measurement; $\sigma^{(j)}$ represents the standard deviation of the meter measurement errors associated with the jth measurement. In the compact vector form, $\mathbf{\eta}$ represents the $m \times 1$ vector of random errors, where m represents the number of measurements in the expanded measurement vector; the weighting matrix W represents the diagonal matrix of the squared reciprocals of the standard deviation values, that is, $diag((1/\sigma^{(j)})^2)$.

The solution to the weighted least squares optimization problem in (4) is obtained using the Gauss-Newton iterative algorithm [7, 8], shown in (6) below.

$$\mathbf{x}^{v+1}(t, t_m) = \mathbf{x}^v(t, t_m) + (\mathbf{H}^T \cdot \mathbf{W} \cdot \mathbf{H})^{-1} \cdot \mathbf{H}^T \cdot \mathbf{W} \cdot \left(\mathbf{z}(t, t_m) - \mathbf{h}(\mathbf{x}^v(t, t_m))\right) \quad (6)$$

In (6), v represents the iteration number, and the algorithm iterates until a predefined resolution ε is attained, i.e., $|\mathbf{x}^{v+1} - \mathbf{x}^v| < \varepsilon$, at which time $\mathbf{x}^{v+1} = \hat{\mathbf{x}}(t, t_m)$ or $\hat{\mathbf{x}}$ - where $\hat{\mathbf{x}}$ represents the state vector of the best-estimate Least-Squares (LS) solution at times t and t_m; H represents the Jacobian of $\mathbf{h}(\hat{\mathbf{x}})$ or $\delta\,\mathbf{h}(\hat{\mathbf{x}}(t, t_m))\Big/\delta\mathbf{x}$ computed at each time-step. The main diagonal of the variance-covariance matrix in (6), that is $(\mathbf{H}^T \cdot \mathbf{W} \cdot \mathbf{H})^{-1}$, provides the variances of the estimated states.

The final step involves the application of the Goodness-of-Fit (GOF) method to mathematically assess how well or *confidently* the model (using the best-estimates least-squares state vector solution to update the state variables) fits the measurements. A high confidence level means the model matches the measurements; a low confidence level indicates the presence of measurement or modeling errors. The mathematical formulation of the GOF statement can be found across [4, 5, 8] and [12].

4 Simulation Setup

4.1 Power System Setup

WinIGS [10] is used to model the system and generate the simulation use-cases. The software also generates the COMTRADE files that are used in a MATLAB script to perform the estimation calculations. The simulated power system single-line diagram is shown in Fig. 4. The main three-phase generator is specified at 13 kV$_{LL}$ and is a positive-sequence sinusoidal 60 Hz voltage source. Two three-phase generator sources are added in-parallel to the main generator to inject 9th (15%) and 11th (10%) order harmonics into the system. The 9th and 11th harmonics have magnitudes of 15% and 10% of the magnitude of the fundamental component, respectively. The phase angles of the 9th and 11th harmonic components are both at zero degrees. The delta-wye (grounded) generator step-up (GSU) power transformer, specified at 14 kV$_{LL}$/125kV$_{LL}$, generates 116 kV$_{LL}$ at the transmission system level. The VT is specified at $115/\sqrt{3}$ kV$_{LN}$:115V. The "LL" and "LN" subscripts denote "Line-to-Line" and "Line-to-Neutral," respectively.

It is worth noting that the 9th or otherwise called *triplen* harmonic voltage generator source is applied across each winding of the GSU transformer and thus electromagneti-cally reproduced on the transmission system side; however, any harmonic generator-side triplen harmonic current ends up circulating within the delta-connected primary winding of the GSU transformer – not passing through, as it classifies as a zero-sequence current.

Therefore, the periodic and continuous-time signal $v(t)$ at the VT primary features a *Total Harmonic Distortion* (THD) value as computed in (8), where the *Fourier Series* decomposition of $v(t)$ is provided in (7).

$$v(t) = V_{average} + \sum_{n=1}^{9,11} V_n \cdot sin(n \cdot w_1 \cdot t + n \cdot \theta_n) \quad (7)$$

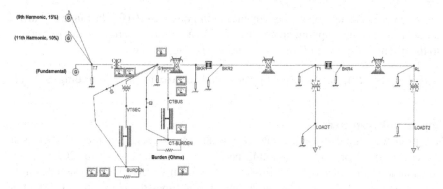

Fig. 4. The simulated test power system.

$$\text{THD}(\%) = \frac{\sqrt{IHC_2^2 + IHC_3^2 + \ldots + IHC_n^2}}{IHC_1} = \frac{\sqrt{IHC_9^2 + IHC_{11}^2}}{IHC_1} = \frac{\sqrt{\{0.15\}^2 + \{0.10\}^2}}{1} = 0.1803 \cong 18\%$$

(8)

where "IHC" means individual harmonic contribution; n indexes the harmonic contributions; V_n is the amplitude of the nth harmonic; w_1 is the radian frequency of the first harmonic or the fundamental signal component; θ_n is the phase shift angle of the nth harmonic, and $V_{average}$ represents any DC or constant term in the signal (typically equaling zero if no nonzero DC term is present).

Note that the energizations of the two power distribution transformers downstream in Fig. 4 inject even-harmonics (particularly second and/or fourth) into the power system. The prominence of each injected harmonic in the frequency spectrum will depend on its individual magnitude relative to that of the fundamental component.

4.2 Sampling Process, Simulation Parameters and Simulation Use-Case Scenarios

Sampling Process

The discrete-time (DT) signal is formed from periodically sampling the analog, continuous-time signal. The continuous-time signal is first operated on using the Shah function $III(x)$ [9], modified by the desired sampling period (Ts), to form the first stage of the sampled signal – that is, a periodic sequence of impulses interspaced in time by Ts, where $fs = 1/Ts$ represents the sampling rate. The sampled signal is then converted from a sequence of impulses to one of sample values, or a discrete sequence, such that each sample in the sequence is located by an integer multiple of Ts. If v(t) is the continuous-time signal, then the periodic sequence of impulses is formed as follows in (9) and (10).

$$III(t) = \sum_{n=-\infty}^{\infty} \delta(t - n \cdot T_s)$$

(9)

$$v_s(t) = III(t) \cdot v(t) = \left\{\sum_{n=-\infty}^{\infty} \delta(t - n \cdot T_s)\right\} \cdot v(t) = \sum_{n=-\infty}^{\infty} v(n \cdot T_s) \cdot \delta(t - n \cdot T_s)$$

(10)

Using $v(n \cdot Ts)$ to abbreviate the sequence of impulses in (10) for the range of n integer indices, a one-to-one mapping process can then be used to convert the series of impulses into the discrete sequence of sample values $v[n]$ – thus, given $v(t)$,

$$v[n] = v(n \cdot T_s), \quad n \, \varepsilon \, Z \quad \text{(set of positive and negative integers, including 0)} \quad (11)$$

Simulation Parameters

The simulated signal sampling rate (f_s) is 80 samples/cycle, which is 4800 samples/second and represents a sampling period or time-interval (Ts) of 208.33 μsecs. The simulation duration is 500 milli-seconds, and thus there are 2400 samples in the discrete-time signal $v[n]$. The quadratic integration time-step (h), also called the DSE time-step, used in the algorithm is twice the sampling period (that is, 416.66 μs).

Simulation Use-Case Scenarios

The VT load is 50 Ω, and the instrumentation channel cable type is Cu-#10. The VT secondary burden comprises the cable leads and the connected load. Three simulations are conducted – the first at a cable length 0.01 miles (52.8 ft); the second at a cable length of 0.08 miles (422.4 ft), and the third at a cable length of 0.16 miles (844.8 ft). Reducing the load resistance and/or increasing the cable length increases the magnitude and phase errors generated in the instrumentation channel. In each simulation use-case, the VT primary and load voltages are measured.

5 Analysis Methods and Presentation of Results

The metered VT secondary voltage, referred to the primary side using the voltage transformation ratio (VTR) of the VT, serves as the uncorrected voltage reference. The metered system voltage serves as the actual primary or ratio voltage reference. The uncorrected voltage reference and the ratio voltage reference are compared to show the errors resulting from the instrumentation channel.

The estimated primary voltage is then compared against the ratio voltage reference to confirm and quantify the degree of error correction. The comparisons are conducted in the time, phasor, and frequency domains.

5.1 Time Domain Comparison Methods

Equation (12) is applied to plot the instantaneous differences between the actual and estimated primary voltages. The normalized-root-mean-square-error (NRMSE) formula, shown in (13), is used to evaluate the estimation residual between the actual and estimated VT primary voltages – where lower residuals indicate less estimation error. In both equations, note that $t = n \cdot Ts$ and n is the sample value index

$$\Delta v(t)_{PER\_UNIT} = \frac{v_p(t) - v_{p\_est}(t)}{v_p(t)} \quad (12)$$

$$\varepsilon_{NRMSE}\% = \frac{\sqrt{\frac{1}{N} \cdot \sum\limits_{j=1}^{N} \left(v_p(j) - v_{p\_est}(j)\right)^2}}{\max(v_p) - \min(v_p)} \times 100 \qquad (13)$$

where N represents the number of sample intervals; v_{p_est} represents the estimated VT primary voltage; v_p represents the actual VT primary voltage.

5.2 Phasor Domain Comparison Methods

The RMS magnitudes of the estimated and uncorrected signals are respectively compared against the RMS magnitude of the ratio voltage reference signal. The phase angles of the estimated and uncorrected signals are also respectively compared against the phase angles of the ratio voltage reference signal. The phase angle and RMS magnitude are derived from the computed phasor. The phasor may be computed over a full, half or quarter cycle. The full-cycle phasor computation approach is taken in this paper.

Generally, consider the **Fourier series** expansion of a cosine sinusoid (14) up to the mth harmonic, and assuming a zero DC term.

$$v(t) = A_1 \cdot \cos(w \cdot t + \phi_1) + A_2 \cdot \cos(2 \cdot w \cdot t + \phi_2) + \ldots + A_m \cos(m \cdot w \cdot t + \phi_m) \qquad (14)$$

where A_m is the maximum value of the mth harmonic; w is the radian frequency and ϕ_m is the phase shift of the mth harmonic. Focusing on the fundamental signal component, (14) can be resolved into (15), where a_1 and a_2 can then be applied in (16) and (17) to form the phasor ($\frac{A_1}{\sqrt{2}} \cdot e^{j\varnothing_1} = A_{1,RMS} \cdot e^{j\varnothing_1}$).

$$v(t_i) = \{a_1\} \cdot \cos(w \cdot t_i) - \{a_2\} \cdot \sin(w \cdot t_i) = \{a_1\} \cdot \cos(w \cdot t_i) + \{-a_2\} \cdot \sin(w \cdot t_i); \qquad (15)$$

$$A_{1,RMS} = \frac{A_1}{\sqrt{2}} = \sqrt{\frac{a_1^2 + a_2^2}{2}} \qquad (16)$$

$$\phi_1 = \tan^{-1}\left(\frac{a_2}{a_1}\right) \qquad (17)$$

More details on phasor computation are provided in the appendices (for continuous-time and discrete-time signals). In the case where the original signal contains harmonics, the phasor computed from (16) and (17) becomes the filtered phasor for the fundamental signal component. The magnitude of the aggregate phasor will feature oscillatory behavior because of the harmonics present. The phase angle of the aggregate phasor will feature the cumulative effects of the phase shifts of the individual harmonic contributions.

The phase angle at each instant (time $t = n \cdot Ts$), after the first cycle or period of samples, can be recursively calculated from the computed full-cycle-phasor. However, under unfaulted, normal system conditions, the phase angle computation from only one period may suffice. The phase angle error (PAE) can then be computed using (18).

$$\Delta\theta(t) = \theta_p(t) - \theta_{p\_est}(t) \quad (or, \quad \Delta\theta_{1-cycle} = \theta_{p\_1-cycle} - \theta_{p\_est\_1-cycle}) \qquad (18)$$

5.3 Frequency Domain Comparison Methods

The frequency spectra of the estimated and uncorrected voltage signals are respectively compared against the spectrum of the ratio voltage reference. The respective signals are discrete sequences of finite lengths; thus, the N-Point Discrete Fourier Transform (N-DFT) [11] is used to generate the respective frequency spectra, where N represents the applied length of the sequence data. The N-DFT formula is shown in (19). Larger N values improve resolution in the frequency spectrum.

$$V[k] = \sum_{n=0}^{N-1} v[n] \cdot e^{-j2\pi nk/N}, \quad 0 \le k < N \tag{19}$$

The difference between the estimated voltage or the uncorrected voltage and the ratio voltage reference generates an error signal. The energy in the error signal, with respect to either the estimated or uncorrected voltage, can be evaluated in the frequency spectrum using the DFT version of Parseval's Theorem [7, 9], as shown in (20).

$$error\_energy = \frac{1}{N} \cdot \sum_{k=0}^{N-1} |\Delta V[k]|^2 = \sum_{n=0}^{N-1} |\Delta v[n]|^2, \quad 0 \le k < N; \ \ 0 \le n < N \tag{20}$$

where $\Delta V[k]$ represents the error signal in the frequency domain and $\Delta v[n]$ represents the error signal in the discrete-time domain. The error-energy formula in (20) can be directly related to the NRMSE (%) index shown earlier in (13) using the relationships in (21) and (22) – where $1/D$ is included as a matching or proportionality factor.

$$error\_energy = \frac{1}{N} \cdot \sum_{k=0}^{N-1} |\Delta V[k]|^2 = \sum_{n=0}^{N-1} |\Delta v[n]|^2 = \frac{1}{D} \cdot \left\{ \frac{(\varepsilon_{NRMSE}\%) \cdot \sqrt{N} \cdot M}{100} \right\}^2 \tag{21}$$

$$M = \max(v_{ratio}) - \min(v_{ratio}) \tag{22}$$

5.4 Results Presentation

Table 1 through Table 4 summarize the results. For brevity, only the plots relating to use-case scenario 3 are shown – that is, resistive load of 50 Ω and cable length of 0.16 miles (844.8 ft). For scenario 3, the time domain, filtered phasor domain and frequency domain plots are presented in Figs. 5 through 8 respectively.

It is important to note that VT resistive load values are typically high (e.g. 1000 Ω), representing low Volt-Ampere (VA) power consumption – which is generally true for electronic or microprocessor loads, as opposed to electromechanical loads. Therefore, reducing the resistive load draws higher inductive current through the instrumentation channel, causing phase and magnitude (or voltage-drop) errors. Increasing the channel cable length increases both the resistance and inductance properties of the channel, thereby also contributing to phase and magnitude errors. Therefore, generally, reducing the load resistance and/or increasing the cable length increases error generation in the instrumentation channel.

Table 1. Ratio error comparison (computed using instantaneous values)

BEFORE CORRECTION			AFTER CORRECTION			
Load (Ω)	Cable Length (ft)	Ratio Error NRMSE(%)	Load (Ω)	Cable Length (ft)	Ratio Error NRMSE(%)	% Error Reduction
50	52.8	0.68	50	52.8	0.1	85.29%
50	422.4	0.91	50	422.4	0.12	86.81%
50	844.8	1.3	50	844.8	0.17	86.92%

Table 2. Phase-angle error comparison computed using the filtered phasor for the fundamental (f_0) component.

BEFORE CORRECTION			AFTER CORRECTION			
Load (Ω)	Cable Length (ft)	f_0 Phase Angle Error (deg)	Load (Ω)	Cable Length (ft)	f_0 Phase Angle Error (deg)	% Error Reduction
50	52.8	0.63	50	52.8	0.02	96.83%
50	422.4	0.67	50	422.4	0.02	97.01%
50	844.8	0.7	50	844.8	0.02	97.14%

Table 3. Error signal energy (full sequence)

BEFORE CORRECTION			AFTER CORRECTION			
Load (Ω)	Cable Length (ft)	Error Signal Energy	Load (Ω)	Cable Length (ft)	Error Signal Energy	% Error Reduction
50	52.8	21600	50	52.8	4.58E+03	78.80%
50	422.4	2.76E+05	50	422.4	8.77E+03	96.82%
50	844.8	9.32E+05	50	844.8	1.98E+04	97.88%

Table 4. Error signal energy. Harmonic Band of $[-12, 12]$, to include the $(\pm)$ 1st, 9th & 11th

BEFORE CORRECTION			AFTER CORRECTION			
Load (Ω)	Cable Length (ft)	Error Signal Energy	Load (Ω)	Cable Length (ft)	Error Signal Energy	% Error Reduction
50	52.8	3.60E+04	50	52.8	7.60E+03	78.89%
50	422.4	4.61E+05	50	422.4	1.38E+04	97.01%
50	844.8	1.56E+06	50	844.8	3.03E+04	98.06%

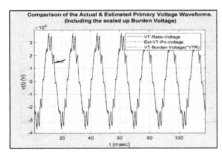

Fig. 5. (Zoom-In) instantaneous plots for use-case scenario 3.

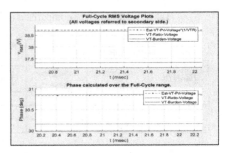

Fig. 6. (Zoom-In) fundamental RMS Ampl. & phase plot for use-case scenario 3.

(a)   **(b)**

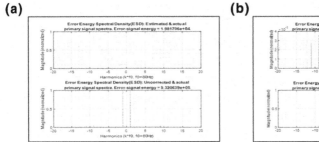

Fig. 7. (a) Full-Spectrum Error Signal ESDs, cross-normalized for easier correlation of energy levels across both plots. (Case 3), (b) (With zoom-in on top plot) Full-spectrum error signal ESDs, cross-normalized for easier correlation of energy levels across both plots. (Case 3)

(a)   **(b)**

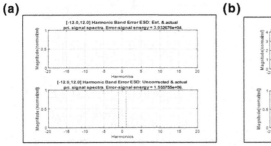
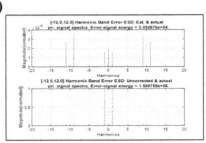

Fig. 8. (a) Error signal ESDs, only computed over the [−12,12] harmonic band, & cross-normalized for easier correlation of energy levels across both plots. (Case 3), (b) (With zoom-in on top plot) Error signal ESDs, only computed over the [−12,12] harmonic band, & cross-normalized for easier correlation of energy levels across both plots. (Case 3)

6 Observations and Conclusion

The results presented in Tables 1, 2, 3, and 4 both confirm and quantify the degrees of correction successfully accomplished by the method under evaluation. Note that in Table 2, the phase angle is computed from the filtered phasor and thus represents the angle of the fundamental frequency component (for either the uncorrected or estimated signals) – here, for each cable length, the error-correction algorithm produces a signal estimate with a fundamental phase angle (FPA) remaining at 99.94% of the FPA of the ratio voltage reference (RVR); while the FPA of the uncorrected voltage is respectively at 97.96%, 97.83% and 97.73% of the FPA of the RVR.

The frequency spectral analyses also confirm successful error correction outcomes by the method under evaluation. Figures 7, 8 show outcomes for the Simulation Use-case Scenario 3, simulating a 50 Ω VT resistive load with a channel cable length of 0.16 miles. Figs. 8a and b focus on a specified harmonic band, while Figs. 7a and b show the full spectrum. It can clearly be seen in Figs. 7 through 8 that the error signal energy spectral density (ESD) at the fundamental frequency is much lower after correction (top subplots) than before correction (bottom subplots) – the same positive observation is made for the 9th and 11th harmonics. Upon zooming in, as shown in Figs. 7b and 8b, it is worth noting that there is some signal energy at the 0th harmonic in the error-signal spectrum after correction, not present before correction. This relatively insignificant but nonetheless present amount of energy at the 0th harmonic (or 0-Hz) indicates the presence of a negligible but nonetheless present DC offset in the estimated time-domain voltage signal – which may result unavoidably from the applied estimation technique, or from minor improvement opportunities with the structure of the error-correction algorithm.

The paper has presented a new method that can correctly estimate the VT primary voltage that is corrupted by harmonic content. The presented method is also algorithmically structured for real-time implementation. Future work includes the development of the error-correction algorithm for the Coupled-Capacitor VT instrumentation channel, in addition to commencing efforts toward the hardware implementation of the method.

7 Appendices

A1: Full-Cycle Phasor Calculation: The Fourier series representation of any given cosine sinusoid is shown in (23), up to the mth harmonic.

$$v(t) = A_1 \cdot \cos(w \cdot t + \phi_1) + A_2 \cdot \cos(2 \cdot w \cdot t + \phi_2) + \ .... \ + A_m \cos(m \cdot w \cdot t + \phi_m)$$

$$(23)$$

The fundamental harmonic sinusoid at time t_i can be simplified using a trigonometry identity, as shown in (24). It can also be decomposed into a linear combination of phase-shifted sinusoids up to a Kth sinusoid, as shown in (25). Equation (25) is re-arranged and further simplified to (26). The expressions in the {.} brackets in (26) respectively represent quadrature constants, a_1 and a_2, shown in (27).

$$v(t_i) = A_1 \cdot \cos(w \cdot t_i + \phi_1) \ = \ \{A_1 \cdot \cos(\phi_1)\} \cdot \cos(w \cdot t_i) \ - \ \{A_1 \cdot \sin(\phi_1)\} \cdot \sin(w \cdot t_i)$$

$$(24)$$

$$v(t_i) = A_1 \cdot \cos(w \cdot t_i + \phi_1)$$

$$= A_{11} \cdot \cos(w \cdot t_i + \phi_{11}) + A_{12} \cdot \cos(w \cdot t_i + \phi_{12}) + \dots + A_{1K} \cdot \cos(w \cdot t_i + \phi_{1K});$$

$$= A_{11} \cdot \{\cos(w \cdot t_i) \cdot \cos(\phi_{11}) - \sin(w \cdot t_i) \cdot \sin(\phi_{11})\} + \dots$$

$$+ A_{1K} \cdot \{\cos(w \cdot t_i) \cdot \cos(\phi_{1K}) - \sin(w \cdot t_i) \cdot \sin(\phi_{1K})\} \tag{25}$$

$$= \{\sum_{J=1}^{K} A_{1K} \cdot \cos(\phi_{1K})\} \cdot \cos(w \cdot t_i) - \{\sum_{J=1}^{K} A_{1K} \cdot \sin(\phi_{1K})\} \cdot \sin(w \cdot t_i); \tag{26}$$

$$= \{a_1\} \cdot \cos(w \cdot t_i) - \{a_2\} \cdot \sin(w \cdot t_i);$$

$$= \{a_1\} \cdot \cos(w \cdot t_i) + \{-a_2\} \cdot \sin(w \cdot t_i); \tag{27}$$

The coefficients of Eqs. (27) and (24) can be respectively compared and mathematically operated upon to derive the phasor magnitude and angle, as shown in (28) through (32).

$$a_1 = A_1 \cdot \cos(\phi_1) = \{\sum_{J=1}^{K} A_{1K} \cdot \cos(\phi_{1K})\} \tag{28}$$

$$a_2 = A_1 \cdot \sin(\phi_1) = \{\sum_{J=1}^{K} A_{1K} \cdot \sin(\phi_{1K})\} \tag{29}$$

$$A_1 = \sqrt{A_1^2 \cdot \{\cos^2(\phi_1) + \sin^2(\phi_1)\}} = \sqrt{a_1^2 + a_2^2} \tag{30}$$

$$A_{1,RMS} = \frac{A_1}{\sqrt{2}} = \sqrt{\frac{a_1^2 + a_2^2}{2}} \tag{31}$$

$$\phi_1 = \tan^{-1}\left(\frac{a_2}{a_1}\right) \tag{32}$$

Now, assume a sinusoidal waveform at the fundamental frequency, and uniformly sampled at N samples per period – where t_i in subsequent equations now means $i \cdot T_S$ or the ith time interval, T_S. The phasor calculation begins by decomposing $v[i] = v(i \cdot T_S)$ into quadrature components A_i and B_i, and respectively summing these values over a period to obtain A and B – shown in (33) and (35). Applying (27), A and B can be respectively resolved and directly related to a_1 in (34) and a_2 in (36). Now, introducing two RMS quadrature parameters, P_1 and P_2, and respectively relating them with a_1 and a_2, as shown in (37) and (38), the full-cycle phasor can be calculated in (39) through (43), and shown to relate to the results in (30 and 31) through (32). The phasor is computed once after each cycle of the sampled signal.

$$A = \sum_{i=1}^{N} v[i] \cdot \cos(w \cdot i \cdot T_S)$$

$$= \{a_1\} \cdot \sum_{i=1}^{N} \cos^2(w \cdot i \cdot T_S) + \{-a_2\} \cdot \sum_{i=1}^{N} \sin(w \cdot i \cdot T_S) \cdot \cos(w \cdot i \cdot T_S);$$

$$= \{a_1\} \cdot \sum_{i=1}^{N} \{\frac{1}{2} + \frac{\cos(2 \cdot w \cdot i \cdot T_S)}{2}\} + \{-a_2\} \cdot \sum_{i=1}^{N} \sin(w \cdot i \cdot T_S) \cdot \cos(w \cdot i \cdot T_S);$$

$$\tag{33}$$

$$= \{a_1\} \cdot \sum_{i=1}^{N} \{\frac{1}{2}\} + 0 + \{-a_2\} \cdot 0 = a_1 \cdot \frac{N}{2}; \tag{34}$$

$$B = \sum_{i=1}^{N} v[i] \cdot \sin(w \cdot i \cdot T_S)$$

$$= \{a_1\} \cdot \sum_{i=1}^{N} \sin(w \cdot i \cdot T_S) \cdot \cos(w \cdot i \cdot T_S) + \{-a_2\} \cdot \sum_{i=1}^{N} \sin^2(w \cdot i \cdot T_S);$$

$$= \{a_1\} \cdot \sum_{i=1}^{N} \sin(w \cdot i \cdot T_S) \cdot \cos(w \cdot i \cdot T_S) + \{-a_2\} \cdot \sum_{i=1}^{N} \{\frac{1}{2} - \frac{\cos(2 \cdot w \cdot i \cdot T_S)}{2}\};$$

$$\tag{35}$$

$$= \{a_1\} \cdot 0 + \{-a_2\} \cdot \sum_{i=1}^{N} \{\frac{1}{2}\} = -a_2 \cdot \frac{N}{2}; \tag{36}$$

$$A = a_1 \cdot \frac{N}{2} \rightarrow a_1 = \frac{2 \cdot A}{N} = \sqrt{2} \cdot P_1 \tag{37}$$

$$B = -a_2 \cdot \frac{N}{2} \rightarrow a_2 = \frac{-2 \cdot B}{N} = \sqrt{2} \cdot P_2 \tag{38}$$

$$P_1 = \frac{a_1}{\sqrt{2}} = \frac{2 \cdot A}{\sqrt{2} \cdot N} = \sqrt{2} \cdot \frac{A}{N} \tag{39}$$

$$P_2 = \frac{a_2}{\sqrt{2}} = \frac{-2 \cdot B}{\sqrt{2} \cdot N} = (-1) \cdot \sqrt{2} \cdot \frac{B}{N} \tag{40}$$

$$P = P_1 + jP_2 = \cdot \frac{\sqrt{2}}{N} \cdot \{A - jB\} \tag{41}$$

$$|P| = \sqrt{P_1^2 + P_2^2} = \frac{\sqrt{2}}{N} \cdot \{\sqrt{A^2 + (-B)^2}\}$$

$$= \sqrt{\frac{a_1^2 + a_2^2}{2}} = \frac{A_1}{\sqrt{2}} = A_{1,RMS} \tag{42}$$

$$\phi_1 = \tan^{-1}\left(\frac{P_2}{P_1}\right) = \tan^{-1}\{\frac{(-1) \cdot \sqrt{2} \cdot \frac{B}{N}}{\sqrt{2} \cdot \frac{A}{N}}\} = \tan^{-1}\{\frac{-B}{A}\}$$

$$= \tan^{-1}\{\frac{a_2}{a_1}\} \tag{43}$$

A2: Half-Cycle Phasor (Magnitude) Calculation: After each half cycle, (44) and (45) can be used to calculate the RMS value of the sampled sinusoid – where H represents half the number of samples per period or $N/2$; j iterates through every H^{th} set of samples, resulting in the calculation of the RMS value every half cycle.

$$i_{avg,j} = \frac{1}{H} \cdot \sum_{k=((j-1)\cdot H + 1)}^{j\cdot H} |i_{sampled}[k]| \tag{44}$$

$$i_{sampled,RMS,j} = 0.9 \cdot i_{avg,j} \tag{45}$$

A3: Quarter-Cycle Phasor Calculation: After each quarter cycle, two quadrature samples are extracted from the sampled signal. Equations (46) through (48) are used to compute the phasor at system frequency.

$$s_1 = i_{sampled}[j] \tag{46}$$

$$s_2 == i_{sampled}[j - (\frac{N}{4} - 1)] \tag{47}$$

$$A_{1,RMS} = \frac{1}{\sqrt{2}} \cdot \sqrt{s_1^2 + s_2^2} \tag{48}$$

References

1. Izykowski, J., Kasztenny, B., Rosolowski, E., Saha, M.M., Hillstrom, B.: Dynamic compensation of capacitive voltage transformers. IEEE Trans. Power Deliv. **13**(1), 116–122 (1998)
2. Hamrita, T., Heck, B., Meliopoulos, S.: On-line correction of errors introduced by instrument transformers in transmission-level steady-state waveform measurements. IEEE Trans. Power Deliv. **15**(4), 1116–1120 (2000)
3. Sun, J., Wen, X., Lan, L., Li, X.: Steady-state error analysis and digital correction for capacitor voltage transformers. In: 2008 International Conference on Electrical Machines and Systems, Wuhan (2008)
4. Obikwelu, C., Meliopoulos, S.: VT instrumentation channel error correction using dynamic state estimation. In: 2021 North American Power Symposium (NAPS), Arizona (2021). (Accepted)
5. Obikwelu, C., Meliopoulos, S.: CT instrumentation channel error correction using dynamic state estimation. In: 2019 North American Power Symposium (NAPS), Wichita (2019)
6. Cokkinides, M.G., Stefopoulos, G.: Quadratic integration method. In: Proceeding of International Power System Transients conference, Montreal (2005)
7. Moon, T., Stirling, W.: Mathematical Methods and Algorithms for Signal Processing. Prentice Hall (2000)
8. Kezunovic, M., Meliopoulos, S., Venkatasubramanian, V., Vittal, V.: Application of Time-Synchronized Measurements in Power System Transmission Networks. Springer, Heidelberg (2014). https://doi.org/10.1007/978-3-319-06218-1
9. Bracewell, R.N., Bracewell, R.N.: The Fourier Transform and its Applications. McGraw-Hill, New York (1986)

10. https://www.ap-concepts.com/win_igs.htm. Advanced Power Concepts
11. Rebizant, W., Szafran, J., Wiszniewski, A.: Digital Signal Processing in Power System Protection and Control. Springer-Verlag, London (2011). https://doi.org/10.1007/978-0-85729-802-7
12. Obikwelu, C., Meliopoulos, S.: CT saturation error correction within merging units using dynamic state estimation. In: 2020 IEEE 3rd International Conference on Renewable Energy and Power Engineering (REPE), Edmonton, Canada, October 2020

Electric Mobility; Renewable Energy

A Single-Phase Current-Source Converter Combined with a Hybrid Converter for Interfacing an Electric Vehicle and a Renewable Energy Source

Catia F. Oliveira[✉], Ana M. C. Rodrigues, Delfim Pedrosa, Joao L. Afonso, and Vitor Monteiro

ALGORITMI Research Centre, University of Minho, Guimarães, Portugal
{c.oliveira,arodrigues,dpedrosa,jla,vmonteiro}@dei.uminho.pt

Abstract. This paper presents a single-phase current-source converter (CSC) combined with a hybrid converter on the dc-link, allowing to interface an electric vehicle (EV) and a renewable energy source (RES). Therefore, the interface with the power grid is only performed through the CSC, which also permits the operation as shunt active power filter (SAPF), allowing to compensate power quality problems related with current and low power factor in the electrical installation. The whole system is composed by two main power stages, namely, the CSC that is responsible for compensating the current harmonics and low power factor, as well as operating as a grid-tied inverter or as an active rectifier, and the hybrid converter that is responsible for interfacing the dc-link of the CSC with the converters for the EV and the RES interfaces. As demonstrated along the paper, the CSC, combined with the hybrid converter on the dc-link, allows the operation as SAPF, as well as the operation in bidirectional mode, specifically for the EV operation, and also for injecting power from the RES. In the paper, the power electronics structure is described and the principle of operation is introduced, supported by the description of the control algorithms. The validation results show the proper operation of the CSC, combined with the hybrid converter on the dc-link, for the main conditions of operation, namely exchanging power with the power grid in bidirectional mode and operating as a SAPF.

Keywords: Current-source converter · Electric vehicle · Hybrid converter · Renewable energy source · Power quality

1 Introduction

Currently, the increasing greenhouse gas emissions caused by conventional vehicles and the demand for electricity is a huge concern. In order to contribute for a more sustainable mobility, the electric vehicle (EV) has been seen as an efficient solution. Moreover, the EV can also be managed for storing and delivering power during the intermittence of the renewable energy sources (RES) [1]. Thus, the EV can be used as an energy storage

J. L. Afonso et al. (Eds.): SESC 2020, LNICST 375, pp. 175–186, 2021.
https://doi.org/10.1007/978-3-030-73585-2_11

system (ESS), permitting to store the energy in periods where the RES production is excessive and in periods of intermittence, delivering the energy back to the installation contributing to balance the power management and consumption [2]. The integration of the EV and RES in the power grid can introduce power quality problems, such as current harmonics and low power factor, due to the required power converters to interface the power grid [3–5]. The interface of the EV and RES with the power grid is performed by an ac-dc converter and by a dc-dc converter. The dc-dc converter ensures a controllable voltage and current of the dc interfaces, whereas the ac-dc converter is responsible to guarantee sinusoidal currents with low harmonic distortion and unitary power factor [6]. Depending on the dc-link, the ac-dc converter can be voltage-source converter (VSC) or current-source converter (CSC). The CSC presents as main advantages a good current control and a short-circuit protection capability. On the other hand, the dc-link of the CSC is composed by an inductor with a longer lifetime comparing with the type of capacitors used in the dc-link of the VSC. Due to these advantages, the CSC can be applied in some power electronic applications, such as in the interface of RES and EV, active power filters and motor drives [7]. Nevertheless, in order to maintain an acceptable level of current ripple in dc-link, it is required a large dc-link inductor for the CSC [8, 9]. In order to reduce the inductance value and consequently the size and costs of this inductor, a hybrid converter is proposed in [10] for a four-wire CSC operating as a shunt active power filter (SAPF), whose structure can also be applied for a single-phase CSC controlled as SAPF, as shown in [11]. Taking into account the perspective of controlling the EV in bidirectional mode for a smart grid perspective, the interface of a conventional CSC with EV is more difficult due to the fact that the CSC is unidirectional. However, the CSC combined with a hybrid converter on the dc-link, can also be used for EV applications, as it is presented in [12]. Additionally, it can also be applied as motor driver in EV applications, as shown in [13]. Figure 1 presents a block diagram showing the CSC combined with a hybrid converter for interfacing an EV and a RES. Regarding the EV charging, several controls can be considered aimed to preserve the batteries life cycle, such as the constant current and constant voltage (CC/CV) due to its simplicity and easy implementation [14–16]. On the other hand, regarding the RES interface, also several control algorithms can be considered, such as the maximum power point tracking (MPPT) based on the perturb and observe strategy [17, 18].

This paper presents a single-phase CSC combined with a hybrid converter for interfacing an EV and a RES. Based on the power electronics structure, the interface with the power grid is only performed through the CSC, and in terms of operation modes, it is possible: (i) operation as SAPF; (ii) operation as EV charger (the EV is charged with power from the power grid); (iii) operation as EV discharger, where part of the stored power in the EV is injected into the power grid; (iv) operation as RES interface, injecting power into the power grid; (v) combined operation of the previous modes (for instance, charging the EV and operating as a SAPF at the same time).

In Sect. 2 is described in detail the topology and in Sect. 3 is presented the control algorithm for the single-phase CSC and the hybrid converter, as well as the compensation control theory. In Sect. 4 are presented simulations results of the single-phase CSC. Finally, Sect. 5 exposes the main conclusions.

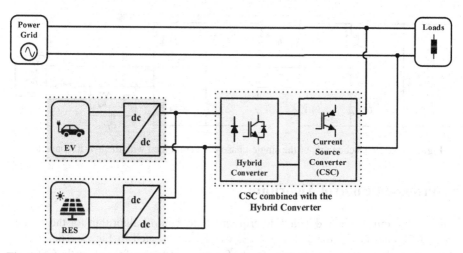

Fig. 1. Block diagram of the CSC combined with the hybrid converter on the dc-link, for interfacing an EV and a RES with the power grid.

2 Principle of Operation

In this section is described the topology presented in Fig. 2. As it can be seen, the topology consists of a CSC and a hybrid converter. The ac-dc converter is required for interfacing the power grid with the dc-link to compensate the current harmonics in the power grid and obtain a unitary power factor. This converter is composed by power semiconductors totally controlled, that it can be MOSFETs, IGBTs or RB-IGBTs. As this converter is unidirectional, when used IGBTs, it is crucial to guarantee block reverse, avoiding that the current flows by the antiparallel of IGBTs. For that reason, the use of RB-IGBTs is a viable solution allowing the reduction of costs, instead of using IGBTs in series with diodes [19, 20]. On the other hand, the hybrid converter is bidirectional and it is composed by two diodes and two IGBTs. The aim of this converter is interfacing the EV, allowing the power absorption or injection in the power grid (charging or discharging the EV) and interfacing the RES (extracting the maximum power) controlling the dc current in dc-link of the CSC. Thus, the hybrid converter has two modes of operation: (i) grid-to-vehicle (G2V) and (ii) vehicle-to-grid (V2G). During the G2V mode, the power is transferred from the power grid to the dc-link of the hybrid converter, where the dc-dc converter for the EV is connected. In this mode of operation is applied the CC/CV method, whose output of the control is a voltage reference for the dc-link of the hybrid converter. During the V2G mode, the power is transferred from the batteries to the power grid. For this operation mode, the output of the control is also a voltage reference applied to the dc-link of the hybrid converter. For both modes of operation, if the dc-link voltage (v_{ev}) is higher than the voltage reference, the IGBTs S_5 and S_6 are on, whereas if v_{ev} is lower than the voltage reference both IGBTs are off. When v_{ev} is equal to the voltage reference, it is only enabled one of the IGBTs, S_5 or S_6.

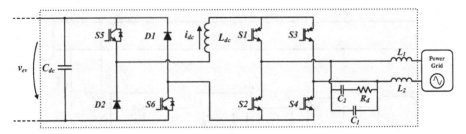

Fig. 2. Electrical schematic of the single-phase CSC combined with the hybrid converter.

3 Proposed Control Algorithm

This section introduces a detailed description of the control algorithm for the single-phase CSC interfacing the EV and RES, which is presented in Fig. 3. The control algorithm includes the control theory for the SAPF operation performed by the CSC, the charging of the EV using the CC/CV method and the extraction of the maximum point of the power from the RES through the MPPT control algorithm. The management of the power from/for the EV and the RES is made by the hybrid converter.

The operation of the SAPF is realized before and during the charging and discharging of the EV and the extraction of the power from the RES. For the correct operation of the SAPF, it is required a method of synchronization with the power grid. In this case, it is used the E-PLL algorithm [21–23] whose input signal is the power grid voltage (v_g). The dc-link regulation (p_{reg}) is obtained with a PI controller, where the input signals are the dc-link current (i_{dc}) and its reference (i_{dc}^*). Based on the E-PLL (v_{pll}) and the load current (i_L), it is calculated the compensation current (i_c) through the theory of Fryze-Buchholz-Depenbrock (FBD) approached in [24–26]. This control theory presents as the main advantage a simpler implementation comparing to others control theories that can be applied to SAPF, for example the p-q theory. The commutation of the power semiconductors of the CSC is obtained from the modulation block of the SAPF. The hybrid converter is controlled for the EV charging following the two stages of the CC/CV method and for the EV discharging. As it can be observed, the variables needed to control the hybrid converter are the EV current (i_{ev}), the EV reference current (i_{ev}^*) and the EV voltage (v_{ev}). Moreover, it is measured the voltage (v_{RES}) and the current (i_{RES}) on RES for the calculation of the power provided by the RES and thus, compare the power obtained with the power calculated in the previous time in order to extract the maximum power and inject power in the power grid. In the following items are described with more detail the power theories used for both converters.

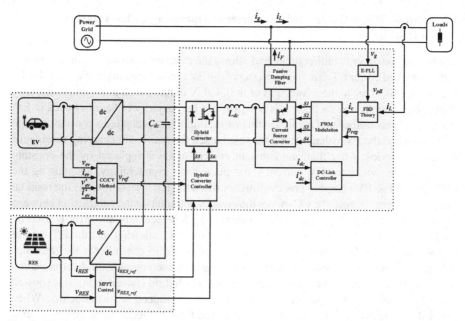

Fig. 3. Block diagram of the control algorithm for the CSC combined with the hybrid converter on the dc-link, interfacing an EV and a RES with the power grid.

3.1 Control When the CSC is Operating as SAPF

The SAPF is responsible for supplying the harmonics and reactive power required by the loads. For that, the SAPF must be capable of producing the compensation current in order to obtain a power grid current sinusoidal and in phase with the power grid voltage. The compensation current is calculated through a power theory, that in this case is the FBD theory. The principle of this theory is that a load can be represented by an equivalent conductance (G_a) in parallel with a current-source. In order to apply this power theory, it is calculated the active power multiplying the output of the E-PLL (v_{pll}) by the load current (i_L), calculating subsequently, the average active power (P) through the sliding window method. The G_a is obtained through the P and the squared RMS value of the power grid voltage (v_g^2), as it can be seen in (1).

$$G_a = \frac{P}{v_g^2}. \tag{1}$$

The active current (i_a) associated to the G_a is presented in (2).

$$i_a = G_a v_g. \tag{2}$$

Finally, the compensation current (i_c) produced by the SAPF is determined by (3), which corresponds to the difference between the load current (i_L) and the active current (i_a).

$$i_c = i_L - i_a. \tag{3}$$

3.2 Control When the Hybrid Converter is Operating as EV Charger or Discharger

The hybrid converter is bidirectional and allows the charging and discharging of the EV with controlled current. The dc-dc converter of the EV is connected in parallel with the dc-link of the CSC. Due to the advantages of the CC/CV method such as its simplicity, easy implementation and the fact that it is adequate for the majority of batteries used in EV, this method was studied and applied in this work. Initially, it is applied a constant current for the EV and the EV voltage increases since reach the maximum charging voltage for the used batteries. After that, the constant current mode is replaced by the constant voltage mode, where the EV current starts decreasing exponentially, maintaining the EV voltage. The EV charging process finishes when the EV current reaches the residual batteries current. During the CC mode, the error is obtained by the subtraction between the reference current (i_{ev}^*) and the EV current (i_{ev}). Subsequently, the error is used by the PI controller, where it is adjusted the gains, kp and ki, allowing that the EV current follows the established reference (i_{ev}^*). The output of the PI controller is the reference voltage (v_{ev}^*) which is compared with EV voltage, whose comparison determines the states of the power semiconductors that composes the hybrid converter. In (4) is showed the conditionals for the commutations of the aforementioned semiconductors. When reached the maximum EV voltage, it is initialized the CV mode whose control is also carried out by the PI controller. In this case, the control variable is the EV voltage instead of the EV current, as previously described for the CC mode.

$$\begin{cases} \text{if } v_{ev} < v_{ev}^*, \ S_5 = 0 \land S_6 = 0 \\ \text{if } v_{ev} > v_{ev}^*, \ S_5 = 1 \land S_6 = 1 \, . \\ \text{if } v_{ev} = v_{ev}^*, \ S_5 = 0 \land S_6 = 1 \end{cases} \qquad (4)$$

3.3 Control When the Hybrid Converter is Operating as RES Interface

Another application of the hybrid converter is the interface of the RES in order to extract the maximum power and to inject in the power grid and/or charging the EV. For that, it is necessary to implement a control algorithm designated MPPT. The MPPT algorithm aims to identify constantly the maximum power point of the RES and allows that the hybrid converter operates correctly. There are several MPPT control algorithms which are approached in [17, 27, 28]. The main differences presented in these MPPT algorithms are the costs, efficiency as well as the complexity of implementation.

4   Simulation Results

In this section are presented the main simulation results of the topology simulated with the PSIM software. The presented simulations were performed for the main conditions of operation of the CSC and the hybrid converter, specifically considering the EV charging and discharging processes and including the compensation of the current harmonics and power factor in the power grid. Table 1 shows the specifications of the developed simulation model.

Initially, the hybrid converter had as main contribution to reduce the inductance value of the inductor in dc-link of the CSC. However, the following results prove that the same converter apart from allowing the reduction of the inductor in dc-link, the hybrid converter is also advantageous for interface other dc systems, like the interface of an EV interface and the interface of a RES. In order to validate the CSC operating as SAPF and the EV charging process, the presented simulation results in this paper consist of: (i) Dc-link current regulation; (ii) Compensation of current harmonics and low power factor in the power grid by the CSC operating as SAPF; (iii) EV charging and discharging along with the harmonic compensation by the CSC operating as SAPF.

As nonlinear load connected to the power grid, it was considered a full-bridge rectifier with a RC load (168 Ω, 2.6 mF), with a coupling inductor of 0.5 mH, as well as a RL load in parallel (12.5 Ω, 70 mH). Table 2 shows the specifications of the dc-link of both converters, including the values of components that composes the passing damping filter. This coupling filter was introduced in order to reduce the losses and the electromagnetic interference caused by the CSC as approached in [29].

Table 1. Specifications of the CSC interfacing EV and RES.

Parameters	Value	Unit
RMS grid voltage	230 ± 10%	V
Grid frequency	50 ± 1%	Hz
Output dc voltage range	0 to 450	V
Switching frequency	40	kHz

Table 2. Specifications of the power converters.

Parameters	Value	Unit
Inductor L_{dc}	100	mH
Capacitor C_{dc}	300	μF
Damping resistor R_d	15	Ω
Inductors L_1, L_2	20	μH
Capacitor C_1	5	μF
Capacitor C_2	20	μF

One of the roles of the CSC is to control and maintain the current in the dc-link at its reference value. The current in dc-link (i_{dc}) is presented in Fig. 4, which is regulated through the PI controller. As it can be observed, when starts the EV charging, i.e., during the transient-state, the dc-link current decreases, however the dc-link current quickly reaches the reference current of 50 A.

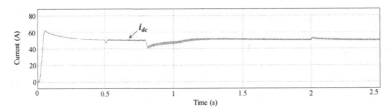

Fig. 4. Simulation results of the dc-link current regulation (i_{dc}).

In Fig. 5 are presented the simulation results of the power grid voltage (v_g) and grid current (i_g) before compensation. As it can be observed, the i_g is not sinusoidal and in phase with the v_g. The i_g presents a THD$_{\%f}$ of 36%. The RMS value of the i_g is 10.28 A. As mentioned above, having a high harmonic distortion in the power grid can result in increasing losses, and in possible damage of equipment connected to the installation.

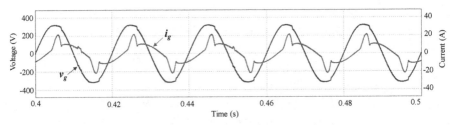

Fig. 5. Simulation results of the power grid voltage (v_g) and grid current (i_g) before compensation, and without the EV charging process and injection of power from RES. The RMS value of the i_g is 10.28 A.

After the dc-link regulation, the SAPF produces a compensation current in order to compensate the current harmonics and low power factor in the power grid. Figure 6 shows v_g and i_g after compensation, where the power grid current is almost sinusoidal and in phase with the power grid voltage. The RMS value of the i_g is 7.35 A. The THD$_{\%f}$ in the power grid current is reduced from 36% to 5.69%, corresponding to an important improvement.

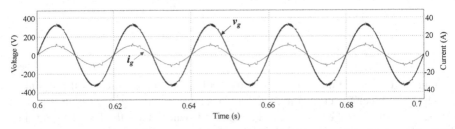

Fig. 6. Simulation results of the power grid voltage (v_g) and grid current (i_g) after compensation, and without the EV charging process and injection of power from RES. The RMS value of the i_g is 7.35 A.

Figure 7 shows the G2V (1 and 2) and V2G (3) modes, where i_{ev} is the EV current and v_{ev} is the EV voltage. During G2V operation mode (from 0.8 s to 2 s), the EV are charged for a reference current of 10 A with a minimum voltage of 250 V. The EV charging process consists of two charging stages, namely constant current (1) followed by constant voltage (2). Initially, the EV is charged with a current of 10 A, whose power is supplied by the power grid and the RES. When the voltage in the EV reaches the maximum voltage of 450 V, the EV current starts decreasing until reaches approximately 0 A (constant voltage). This method is very advantageous for the charging of lithium batteries, that are very common in EV. During V2G mode (from 2 to 2.5 s), the power from the EV is injected into the power grid for a reference current of 2 A, which corresponds to the stage (3). As it can be observed during the discharging of the EV, the v_{ev} decreases. During all modes (1, 2 and 3), the CSC is controlled to operates as SAPF, i.e., compensates the current harmonics in the power grid and to obtain an unitary power factor.

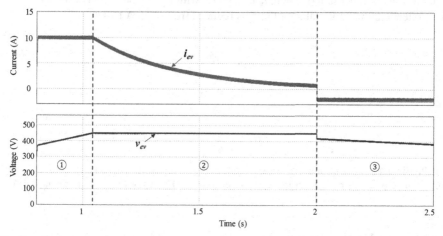

Fig. 7. Simulation results of a Lithium battery charging and discharging process: (1) CC stage; (2) CV stage and (3) EV discharging.

During the EV charging and injection of power in the power grid, it cannot be neglected the power quality improvement. So, the SAPF continuing operating and the current harmonics in power grid is compensated. In Fig. 8 are showed v_g and i_g after compensation during EV charging process. The power grid current presents a low THD$_{\%f}$ of 2.48%, which is very significantly reduced, as it is intended. On the other hand, the RMS value of the i_g increases to 26.7 A.

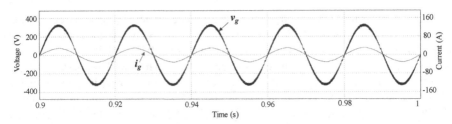

Fig. 8. Simulation results of the power grid voltage (v_g) and grid current (i_g) after compensation, and during the EV charging process without injection of power from RES. The RMS value of the i_g is 26.7 A.

Finally, it is carried out the injection of power from RES, where it is also effectuated the compensation of the current harmonics in the power grid side. Figure 9 presents the v_g and i_g after compensation, where i_g is in phase with v_g, whose THD$_{\%f}$ value is 7.91%. As it intended, the RMS value of the i_g is reduced from 7.35 A to 4 A.

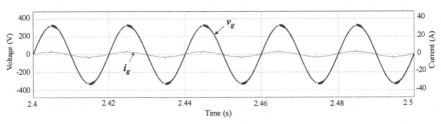

Fig. 9. Simulation results of the power grid voltage (v_g) and grid current (i_g) after compensation, and during the injection of power in the power grid from RES. The RMS value of the i_g is 4 A.

5   Conclusions

This paper presents a single-phase current-source converter (CSC) with a hybrid converter on the dc-link to interface an electric vehicle (EV) and a renewable energy source (RES). The CSC can be controlled as shunt active power filter (SAPF) for compensating current harmonics and low power factor, and also to allow the bidirectional operation for the EV and RES interface. With the objective of preserving the batteries lifetime, the constant current and constant voltage (CC/CV) control was adopted. Regarding the RES interface, it was implemented a maximum power point tracking algorithm. The topology and the operation principles are described along the paper, and simulation results are presented. The obtained results show the correct operation of the CSC operating as SAPF, as well as operating in bidirectional mode to interface the EV and the RES.

Acknowledgment. This work has been supported by FCT – Fundação para a Ciência e Tecnologia with-in the Project Scope: UIDB/00319/2020. This work has been supported by the FCT Project newERA4GRIDs PTDC/EEI-EEE/30283/2017 and the FCT Project DAIPESEV PTDC/EEI-EEE/30382/2017.

References

1. Afonso, J.L.: Sustainable Energy for Smart Cities: First EAI International Conference, SESC 2019, Braga, Portugal, 4–6 December 2019, Proceedings. Springer, Cham (2020). https://doi.org/10.1007/978-3-030-45694-8
2. Monteiro, V., et al.: Assessment of a battery charger for electric vehicles with reactive power control. In: IECON 2012-38th Annual Conference on IEEE Industrial Electronics Society, pp. 5142–5147, October 2012
3. Strasser, T., et al.: A review of architectures and concepts for intelligence in future electric energy systems. IEEE Trans. Ind. Electron. **62**(4), 2424–2438 (2014)
4. Monteiro, V., Pinto, J.G., Exposto, B., Afonso, J.L.: Comprehensive comparison of a current-source and a voltage-source converter for three-phase EV fast battery chargers. In: Proceedings - 2015 9th International Conference on Compatibility Power Electronics, CPE 2015, no. June, pp. 173–178 (2015)
5. Moses, P.S., Deilami, S., Masoum, A.S., Masoum, M.A.S.: Power quality of smart grids with plug-in electric vehicles considering battery charging profile. In: IEEE PES Innovative Smart Grid Technologies Conference Europe ISGT Europe, pp. 1–7 (2010)
6. Monteiro, V., Pinto, J.G., Exposto, B., Monteiro, L.F.C., Couto, C., Afonso, J.L.: A Novel architecture of a bidirectional bridgeless interleaved converter for EV battery chargers. In: IEEE International Symposium on Industrial Electronics, vol. 2015-September, pp. 190–195 (2015)
7. Exposto, B., Monteiro, V., Pinto, J.G., Pedrosa, D., Meléndez, A.A.N., Afonso, J.L.: Three-phase current-source shunt active power filter with solar photovoltaic grid interface. In: Proceedings of the IEEE International Conference on Industrial Technology, vol. 2015-June, no. June, pp. 1211–1215 (2015)
8. Exposto, B., Pinto, J.G., Monteiro, V., Pedrosa, D., Gonçalves, H., Afonso, J.L.: Experimental and simulation results of a current-source three-phase shunt active power filter using periodic-sampling. In: Annual Seminar on Automation Industrial Electronics and Instrumentation 2012, SAAEI 2012, Guimarães, Portugal, pp. 380–385 (2012)
9. Exposto, B., Pinto, J.G., Pedrosa, D., Monteiro, V., Goncalves, H., Afonso, J.L.: Current-source shunt active power filter with periodic-sampling modulation technique. In: IECON Proceedings (Industrial Electronics Conference), pp. 1274–1279 (2012)
10. Pettersson, S., Salo, M., Tuusa, H.: Optimal DC current control for four-wire current source active power filter. In: Conference Proceedings - IEEE Applied Power Electronics Conference and Exposition - APEC, pp. 1163–1168 (2008)
11. Oliveira, C.F., Barros, L.A.M., Afonso, J.L., Pinto, J.G., Exposto, B., Monteiro, V.: A novel single-phase shunt active power filter based on a current-source converter with reduced dc-link. In: Afonso, J.L., Monteiro, V., Pinto, J.G. (eds.) SESC 2019. LNICSSITE, vol. 315, pp. 269–280. Springer, Cham (2020). https://doi.org/10.1007/978-3-030-45694-8_21
12. Su, G.J., Tang, L.: Current source inverter based traction drive for EV battery charging applications. In: 2011 IEEE Vehicular Power and Propulsion Conference, VPPC 2011, pp. 1–6 (2011)
13. Tang, L., Su, G.J.: Boost mode test of a current-source-inverter-fed permanent magnet synchronous motor drive for automotive applications. In: 2010 IEEE 12th Workshop on Control and Modeling for Power Electronics, COMPEL 2010 (2010)
14. Shen, W., Vo, T.T., Kapoor, A.: Charging algorithms of lithium-ion batteries: an overview. In: Proceedings of the 2012 7th IEEE Conference on Industrial Electronics and Applications, ICIEA 2012, pp. 1567–1572 (2012)
15. Pinto, J.G., et al.: Bidirectional battery charger with Grid-to-Vehicle, Vehicle-to-Grid and Vehicle-to-Home technologies. In: IECON Proceedings (Industrial Electronics Conference), pp. 5934–5939 (2013)

16. Moon, J.S., Lee, J.H., Ha, I.Y., Lee, T.K., Won, C.Y.: An efficient battery charging algorithm based on state-of-charge estimation for electric vehicle. In: 2011 International Conference on Electrical Machines and Systems, ICEMS 2011 (2011)

17. Abdelsalam, A.K., Massoud, A.M., Ahmed, S., Enjeti, P.N.: High-performance adaptive perturb and observe MPPT technique for photovoltaic based microgrids. IEEE Trans. Power Electron. **26**(4), 1010–1021 (2011)

18. Esram, T., Chapman, P.L.: Comparison of photovoltaic array maximum power point tracking techniques. IEEE Trans. Energy Convers. **22**(2), 439–449 (2007)

19. Routimo, M., Salo, M., Tuusa, H.: Comparison of voltage-source and current-source shunt active power filters. IEEE Trans. Power Electron. **22**(2), 636–643 (2007)

20. Salo, M., Pettersson, S.: Current-source active power filter with an optimal DC current control. In: PESC Record - IEEE Annual Power Electronics Specialists Conference (2006)

21. Carneiro, H., Monteiro, L.F.C., Afonso, J.L.: Comparisons between synchronizing circuits to control algorithms for single-phase active converters. In: IECON Proceedings (Industrial Electronics Conference), pp. 3229–3234 (2009)

22. Karimi-Ghartemani, M., Khajehoddin, S.A., Jain, P.K., Bakhshai, A., Mojiri, M.: Addressing DC component in PLL and notch filter algorithms. IEEE Trans. Power Electron. **27**(1), 78–86 (2012)

23. Karimi-Ghartemani, M., Iravani, M.R.: A method for synchronization of power electronic converters in polluted and variable-frequency environments. IEEE Trans. Power Syst. **19**(3), 1263–1270 (2004)

24. Czarnecki, L.S.: Budeanu and Fryze: two frameworks for interpreting power properties of circuits with nonsinusoidal voltages and currents. Electr. Eng. **80**(Teoria de Potência), 359–367 (1997)

25. Zhou, J., Wang, Z., Fu, X.: Study on the improved harmonic detection algorithm based on FBD theory. In: Asia-Pacific Power Energy Engineering Conference on APPEEC, no. 1, pp. 2–5 (2011)

26. Staudt, V.: Fryze - Buchholz - Depenbrock: a time-domain power theory, p. 12 (2008)

27. Sera, D., Teodorescu, R., Hantschel, J., Knoll, M.: Optimized maximum power point tracker for fast-changing environmental conditions. IEEE Trans. Ind. Electron. **55**(7), 2629–2637 (2008)

28. Li, S., Zhang, B., Xu, T., Yang, J.: A new MPPT control method of photovoltaic grid-connected inverter system. In: The 26th Chinese Control and Decision Conference (2014 CCDC), pp. 2753–2757 (2014)

29. Hensgens, N., Silva, M., Oliver, J.A., Cobos, J.A., Skibin, S., Ecklebe, A.: Optimal design of AC EMI filters with damping networks and effect on the system power factor. In: 2012 IEEE Energy Conversion Congress and Exposition ECCE 2012, vol. 7, no. 2, pp. 637–644 (2012)

Battery Charging Station for Electric Vehicles Based on Bipolar DC Power Grid with Grid-to-Vehicle, Vehicle-to-Grid and Vehicle-to-Vehicle Operation Modes

Tiago J. C. Sousa[✉], Vítor Monteiro, Sérgio Coelho, Luís Machado, Delfim Pedrosa, and João L. Afonso

Centro ALGORITMI, University of Minho, Campus de Azurém, Guimarães, Portugal
tsousa@dei.uminho.pt

Abstract. This paper proposes an electric vehicle (EV) battery charging station (EV-BCS) based on a bipolar dc power grid with the capabilities of returning energy back to the power grid (vehicle-to-grid – V2G mode), as well as to perform power transfer between different EVs connected to the EV-BCS without drawing power from the power grid (vehicle-to-vehicle – V2V mode), besides the traditional battery charging operation (grid-to-vehicle – G2V mode). The proposed EV-BCS is modular, using three-level bidirectional dc-dc converters. In this paper, for simplicity reasons, only two converters, and hence two EVs, are considered in order to validate the previously referred operation modes. Furthermore, unbalanced operation from the EVs side is also considered for all the operation modes, aiming to consider a real scenario of operation. Simulation results verify the correct operation of the EV-BCS in all cases, with balanced and unbalanced current consumption from the EVs resulting always in balanced currents from the bipolar dc power grid side.

Keywords: Electric vehicle · Battery charging station · Bipolar dc Power Grid · Three-Level dc-dc Converter

1 Introduction

It is well-known that electric vehicles (EVs) are a promising alternative to the conventional vehicles based on internal combustion engines regarding the emission lessening of greenhouse gasses at the user level, as well as the reduction in the exploitation of fossil resources [1, 2]. Furthermore, in addition to the aforementioned advantages in terms of mobility, EVs are also a promising solution concerning smart grids, being able to render ancillary services in conjunction with renewable energy sources and energy storage systems [3, 4, 5]. Besides the traditional battery charging operation, i.e. grid-to-vehicle (G2V) operation mode, one of the first operation modes proposed in the literature regarding the connection of EVs to the power grid was the vehicle-to-grid (V2G) [6, 7],

J. L. Afonso et al. (Eds.): SESC 2020, LNICST 375, pp. 187–199, 2021.
https://doi.org/10.1007/978-3-030-73585-2_12

consisting of using part of the energy stored in the EV battery to deliver it to the power grid. Other operation modes for the EV in the context of smart grids can be analyzed in [8]. However, to make EVs ascend to a global level, a suitable infrastructure in terms of battery charging is mandatory. In this sense, several scheduling strategies regarding EV battery charging stations (EV-BCSs) have been proposed in the literature [9, 10, 11, 12], as well as battery swapping strategies [13, 14, 15, 16] and the combination with energy storage systems and solar photovoltaic panels [17, 18, 19].

In addition to EVs, dc power grids have also been gaining attention, mainly due to the fact that energy storage systems and solar photovoltaic panels, which are major assets towards distributed generation and, consequently, smart grids and microgrids, operate in dc. Moreover, dc power systems are more efficient than their ac counterparts, not suffering from skin effect, and not presenting electrical issues such as reactive power and harmonic currents. In this sense, dc microgrids have been an important topic of research [20, 21]. Among dc power grids can be found unipolar or bipolar types, whether they are comprised by one active conductor plus neutral or two active conductors plus neutral, respectively. As suggested by their designation, bipolar dc power grids provide two symmetrical voltages referenced to the same potential, i.e., the neutral wire. Additionally, it is possible to obtain a voltage that is the double of the base value, namely by using the negative rail as reference instead of the neutral wire. This is advantageous in the way that allows to have two different voltage values instead of only one, as happens with unipolar dc power grids. Besides, bipolar dc power grids are more reliable and present a higher energy transmission capacity [22, 23, 24].

The application of bipolar dc power grids in EV-BCSs has been already addressed in the literature. In [25] is presented an EV-BCS based on a neutral point clamped front-end ac-dc converter, whose split dc-link generates a bipolar dc power grid, where the dc-dc converters responsible for the battery charging operation are connected. However, only the front-end ac-dc converter is addressed in such work. In [26, 27, 28, 29] is studied the utilization of three-level dc-dc converters in an EV-BCS based on a bipolar dc power grid, but the V2G operation mode is not addressed. In this context, this paper presents an EV-BCS based on a bipolar dc power grid with G2V and V2G capability. Besides, the vehicle-to-vehicle (V2V) operation mode is also taken into consideration, which is a relatively recent operation mode regarding EVs but being already a relevant research topic [30, 31, 32, 33]. It should be noted that the front-end ac-dc converter of the EV-BCS is not addressed in this paper.

The rest of the paper is structured as follows: Sect. 2 presents the structure of the EV-BCS, as well as the control of the power converters; Sect. 3 presents the simulation model and results of the EV-BCS in the different possible operation modes; finally, Sect. 4 finalizes the paper with the main conclusions.

2 Structure of the Electric Vehicle Battery Charging Station

This section describes the power structure of the proposed EV-BCS, namely the dc-dc converter used and its control system. The chosen topology for the dc-dc converter is the three-level two-quadrant buck-boost, since it is bidirectional, allowing bidirectional power flow and, hence, the G2V and V2G operation modes, among others, and also because it has a split dc-link in its high voltage side, making it suitable for bipolar dc power grids [34, 35]. Figure 1 shows the power structure of the EV-BCS, where two three-level two-quadrant buck-boost dc-dc converters can be seen, both sharing the same dc-link, which is connected to a bipolar dc power grid. The figure suggests the modularity of the EV-BCS, being possible to connect an indefinite number of similar dc-dc converters to the bipolar dc power grid. For the sake of simplicity, the analysis carried out in this paper only comprises two dc-dc converters, consequently representing two EVs. In [36] a study of the same power structure can be found, i.e., two three-level two-quadrant buck-boost dc-dc converters connected to each other by the high voltage side, focusing on smart grid applications.

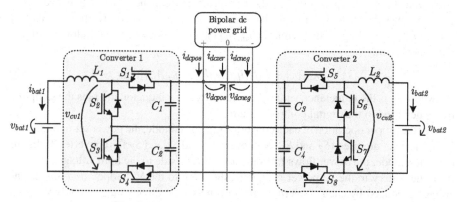

Fig. 1. Power structure of the proposed EV-BCS.

Each converter x (with x varying from 1 to the total number of converters, in this case being 2) is connected to an EVx in the low voltage side, with v_{batx} and i_{batx} being its battery voltage and current, respectively, and to the bipolar dc power grid in the high voltage side, with the positive rail voltage (v_{dcpos}) being applied to the upper capacitor (C_{2x-1}) and the negative rail voltage (v_{dcneg}) being inversely applied to the lower capacitor (C_{2x}). In a bipolar dc power grid, the condition $v_{dcpos} = -v_{dcneg}$ is verified; therefore, the total voltage in the high voltage side of each converter x, i.e., across the series connection of the capacitors C_{2x-1} and C_{2x}, is $2v_{dcpos}$. However, due to the three-level characteristic of the converter, each power semiconductor withstands only a maximum voltage of v_{dcpos}.

For the proper operation of each converter x in buck mode, the power semiconductors S_{4x-3} and S_{4x} are used (S_1 and S_4 for converter 1 and S_5 and S_8 for converter 2). In this operation mode, the power flows from the high voltage side to the low voltage side, i.e., from the dc power grid to the battery, corresponding to the G2V operation mode. The voltage produced by each converter x (v_{cvx}) can assume three values, namely 0, v_{dcpos} and $2v_{dcpos}$. Equation (1) shows the possible values of the voltage v_{cvx} as a function of the switching state of the power semiconductors in buck mode, with 0 meaning the off state and 1 meaning the on state, and Eq. (2) shows the two possible operating regions for the converter in function of the voltages v_{batx} and v_{dcpos} and the duty-cycle (D). It should be noted that the switching signals of the two power semiconductors must be

180° phase shifted in order to assure the proper operation of the converter, also doubling the frequency of the voltage produced by the converter with respect to its switching frequency.

$$v_{cvx} = \begin{cases} 0, S_{4x-3} = 0, S_{4x} = 0 \\ v_{dcpos}, S_{4x-3} = 0(1), S_{4x} = 1(0). \\ 2v_{dcpos}, S_{4x-3} = 1, S_{4x} = 1 \end{cases} \tag{1}$$

$$\begin{aligned} \text{If } v_{cvx} < 2v_{dcpos} - v_{batx} &\rightarrow D < 50\%, v_{cvx} = \{0, v_{dcpos}\}, \\ \text{If } v_{cvx} > 2v_{dcpos} - v_{batx} &\rightarrow D > 50\%, v_{cvx} = \{v_{dcpos}, 2v_{dcpos}\}. \end{aligned} \tag{2}$$

In order to control each EV battery current, a predictive control strategy was employed, based on the model of the converter. Hence, the voltage that each converter x must produce (v_{cvx}) in a given instant k in order to control the EV battery current i_{batx} according to its reference ($i_{batrefx}$) in buck mode obeys the following digital implementation:

$$v_{cvx}[k] = v_{batx}[k] + L_x f_s (i_{batrefx}[k] - i_{batx}[k]), i_{bat} > 0, \tag{3}$$

where L_x is the inductance value of dc-dc converter x inductor and f_s is the sampling frequency used in the digital control system.

For the proper operation of each converter x in boost mode, the power semiconductors S_{4x-2} and S_{4x-1} are used (S_2 and S_3 for converter 1 and S_6 and S_7 for converter 2). In this operation mode, the power flows from the low voltage side to the high voltage side, i.e., from the battery to the dc power grid, corresponding to the V2G operation mode. The voltage produced by each converter x (v_{cvx}) can assume three values, namely 0, v_{dcpos} and $2v_{dcpos}$. Equation (4) shows the possible values of the voltage v_{cvx} as a function of the switching state of the power semiconductors in boost mode, with 0 meaning the off state and 1 meaning the on state, and Eq. (5) shows the two possible operating regions for the converter in function of the voltages v_{batx} and v_{dcpos} and the duty-cycle (D). Similarly to the buck mode, the switching signals of the two power semiconductors must be 180° phase shifted in order to assure the proper operation of the converter, also doubling the frequency of the voltage produced by the converter with respect to its switching frequency.

$$v_{cvx} = \begin{cases} 0, S_{4x-2} = 1, S_{4x-1} = 1 \\ v_{dcpos}, S_{4x-2} = 1(0), S_{4x-1} = 0(1). \\ 2v_{dcpos}, S_{4x-2} = 0, S_{4x-1} = 0 \end{cases} \tag{4}$$

$$\begin{aligned} \text{If } v_{cvx} > 2v_{dcpos} - v_{batx} &\rightarrow D > 50\%, v_{cvx} = \{0, v_{dcpos}\}, \\ \text{If } v_{cvx} < 2v_{dcpos} - v_{batx} &\rightarrow D < 50\%, v_{cvx} = \{v_{dcpos}, 2v_{dcpos}\}. \end{aligned} \tag{5}$$

In order to control each battery current, a predictive control strategy was employed, based on the model of the converter. Hence, the voltage that each converter x must produce (v_{cvx}) in a given instant k in order to control the battery current i_{batx} according to its reference ($i_{batrefx}$) in boost mode obeys the following digital implementation:

$$v_{cvx}[k] = v_{batx}[k] - L_x f_s (i_{batx}[k] - i_{batrefx}[k]), i_{bat} < 0, \tag{6}$$

where L_x is the inductance value of dc-dc converter x inductor and f_s is the sampling frequency used in the digital control system.

3   Computational Simulations

This section presents the simulation parameters and results of the EV-BCS for two EVs, being addressed the G2V, V2G and V2V operation modes, as well as a combination of V2V with G2V and V2V with V2G. The simulations were carried out in the software PSIM v9.1 from Powersim. In Fig. 2 it can be seen the used battery model, namely the Thevenin model, comprised by the open-circuit voltage (v_{ocx}), a capacitor to emulate the dynamic behavior of the battery (C_{batx}), a resistor connected in parallel with the capacitor to emulate the battery self-discharge (R_{px}) and a series resistor, meaning the internal resistance of the battery (R_{sx}). Table 1 presents the parameters of the power converter and batteries of each EV (where it can be seen that the converters are equal), as well as the batteries, which present different initial voltage values, with v_{bat1} starting with 250 V and v_{bat2} with 200 V.

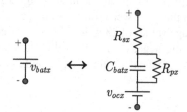

Fig. 2.   Battery model used for each EVx.

Table 1.   Simulation parameters of the EV-BCS and EV batteries.

PARAMETER	VALUE
Initial v_{bat1}	250 V
Initial v_{bat2}	200 V
v_{ocx}	150 V
C_{batx}	0.5 F
R_{sx}	0.1 Ω
R_{px}	100 kΩ
L_x	500 µH
C_{2x-1}, C_{2x}	100 µF
v_{dcpos}	200 V
v_{dcneg}	−200 V
dc power grid impedance	0.1 Ω, 10 µH
Switching frequency	50 kHz
Sampling frequency	50 kHz

Figure 3 shows the normal operation of both EVs at a charging station, i.e., both operating in G2V, and both charging their batteries with the same value of current (20 A).

The figure shows the battery voltages (v_{bat1} and v_{bat2}) and currents (i_{bat1} and i_{bat2}), the currents drawn from the dc power grid, namely in the positive rail (i_{dcpos}), neutral rail (i_{dczer}) and negative rail (i_{dcneg}), and the voltages produced by the converters (v_{cv1} and v_{cv2}). It can be seen that the battery voltages are slightly higher than their original values (2 V higher), which is due to the internal resistance of the batteries and not due to the energy accumulation process, given that the initial instant of figure is 2 ms. It can be seen that both battery currents present the same average value of 20 A, but i_{bat2} presents a much smaller ripple than i_{bat1}. This is due to the fact that the voltage v_{bat2} is practically half (202 V) the total dc power grid voltage, which makes the three-level buck-boost dc-dc converter operate in a region of strong ripple cancelling. This is visible in the voltage produced by this converter (v_{cv2}), presenting a very low duty-cycle between voltage levels 200 V and 400 V (in other words, presenting a duty-cycle slightly higher than 50%). It is noticeable from the voltage v_{cv1} that converter 1 operates with the same voltage levels but with a higher duty-cycle, meaning a higher ripple in i_{bat1}. Regarding the currents absorbed from the dc power grid, it can be perceived that i_{dcpos} and i_{dcneg} are symmetrical, with average values of 23 A and -23 A, respectively, the first one being positive and the second being negative, meaning that the dc power grid is providing power. The current i_{dczer} is the negative sum of i_{dcpos} and i_{dcneg}, therefore presenting a null average value.

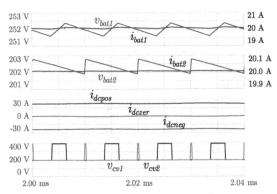

Fig. 3. Simulation results of the G2V operation mode when EV1 and EV2 are charging with a current of 20 A.

Figure 4 shows the operation of both EVs in G2V but with different values of current in order to simulate a case of unbalance. EV1 is charging with a current of 20 A, while EV2 is charging with a current of 40 A. The figure shows the battery voltages (v_{bat1} and v_{bat2}) and currents (i_{bat1} and i_{bat2}), the currents drawn from the dc power grid, namely in the positive rail (i_{dcpos}), neutral rail (i_{dczer}) and negative rail (i_{dcneg}), and the voltages produced by the converters (v_{cv1} and v_{cv2}). In this case, the voltage v_{bat2} presents a value of 204 V, showing the effect of the battery internal resistance when higher currents are applied. It can be seen that both battery currents present the expected average value, with EV1 presenting the same results as the previous case. Despite being a higher current, i_{bat2} still has a low ripple due to the same reason as previously mentioned, as it can be seen from voltages v_{cv1} and v_{cv2}. Regarding the currents absorbed from the dc power grid, i_{dcpos} and i_{dcneg} are symmetrical but with a higher average value than previously (33.6 A), the first one being positive and the second being negative. Thus, the EV-BCS is able to consume balanced currents from the bipolar dc power grid even with unbalanced battery charging operation. Accordingly, the current i_{dczer} presents a null average value.

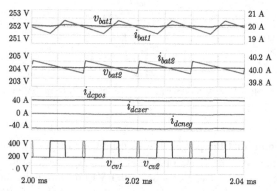

Fig. 4. Simulation results of the G2V operation mode when EV1 is charging with a current of 20 A and EV2 is charging with a current of 40 A.

Figure 5 shows the V2G operation mode for both EVs, discharging their batteries with the same value of current (20 A). This figure shows the battery voltages (v_{bat1} and v_{bat2}) and currents (i_{bat1} and i_{bat2}), the currents drawn from the dc power grid, namely in the positive rail (i_{dcpos}), neutral rail (i_{dczer}) and negative rail (i_{dcneg}), and the voltages produced by the converters (v_{cv1} and v_{cv2}). It can be seen that both battery currents are negative, meaning that the power flow is established from the batteries to the dc power grid, as expected in the V2G operation mode. Also, both battery currents present the same average value of -20 A, but i_{bat2} presents a much smaller ripple than i_{bat1}, which is due to the same reason as aforementioned. In this case, the voltage produced by converter 2 (v_{cv2}) presents a very high duty-cycle between voltage levels 0 V and 200 V (in other words, presenting a duty-cycle slightly smaller than 50%). This happens due to the internal resistance of the batteries, which decreases the battery voltage when current is being supplied by the battery, as it can be seen by the v_{bat2} value of 198 V, which is lower than half the total dc power grid voltage. Regarding the dc power grid currents, it can be seen that i_{dcpos} and i_{dcneg} are symmetrical, but with i_{dcpos} being negative and i_{dcneg} being positive. This means that the dc power grid is receiving power instead of supplying it, as expected from the V2G operation mode. The average value of these currents is 22.1 A, with the current i_{dczer} presenting a null average value.

Figure 6 shows the operation of both EVs in V2G but with different values of current in order to simulate a case of unbalance. EV1 is discharging with a current of 20 A, while EV2 is discharging with a current of 40 A. The figure shows the battery voltages (v_{bat1} and v_{bat2}) and currents (i_{bat1} and i_{bat2}), the currents drawn from the dc power grid, namely in the positive rail (i_{dcpos}), neutral rail (i_{dczer}) and negative rail (i_{dcneg}), and the voltages produced by the converters (v_{cv1} and v_{cv2}). Once again, both battery currents

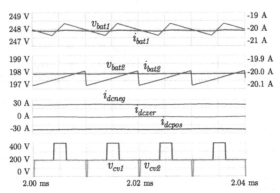

Fig. 5. Simulation results of the V2G operation mode when EV1 and EV2 are discharging with a current of 20 A.

are negative, meaning that the power flow is established from the batteries to the dc power grid, as expected in the V2G operation mode. It is noticeable that both battery currents present the expected average value, with EV1 presenting the same results as the previous scenario. In this case, the voltage v_{bat2} presents a value of 196 V, showing the effect of the battery internal resistance when higher currents are drawn from the battery. Despite being a higher current, i_{bat2} still has a low ripple due to the same reason as previously mentioned, as it can be seen from voltages v_{cv1} and v_{cv2}. Regarding the dc power grid currents, it is noticeable that i_{dcpos} is negative and i_{dcneg} is positive, as in the previous case, meaning that the dc power grid is receiving power instead of supplying it. Moreover, these currents are symmetrical, meaning that the EV-BCS is able to handle unbalances in the power injected by the EVs without unbalancing the dc power grid currents. The average value of these currents is 31.5 A, with the current i_{dczer} presenting a null average value.

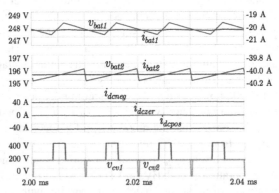

Fig. 6. Simulation results of the V2G operation mode when EV1 is discharging with a current of 20 A and EV2 is discharging with a current of 40 A.

Figure 7 shows the V2V operation mode, where EV1 provides power to EV2. EV2 is charging with a current of 20 A, while EV1 provides the necessary current to perform the battery charging of EV2 without using additional power from the dc power grid. The figure shows the battery voltages (v_{bat1} and v_{bat2}) and currents (i_{bat1} and i_{bat2}), the currents drawn from the dc power grid, namely in the positive rail (i_{dcpos}), neutral rail (i_{dczer}) and negative rail (i_{dcneg}), and the voltages produced by the converters (v_{cv1} and v_{cv2}). In this operation mode, i_{bat1} is negative, similar to V2G, but i_{bat2} is positive, similar to G2V. It can be seen that i_{bat2} has the expected average value of 20 A, with i_{bat1} presenting an average value of approximately −16.3 A. In this case, the voltage v_{bat2} has a value of 202 V, making the produced voltage v_{cv2} alternate between voltage levels 200 V and 400 V. Regarding the dc power grid currents, it can be seen that i_{dcpos} and i_{dcneg} are overlapped and with a practically null average value, meaning that the dc power grid is neither receiving nor providing significant power. The current i_{dczer} presents a similar waveform, also with a null average value, as in the previous cases.

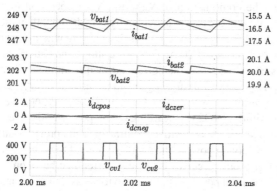

Fig. 7. Simulation results of the V2V operation mode when EV1 is discharging with a current of 16.3 A and EV2 is charging with a current of 20 A.

Figure 8 shows the combination of V2V and G2V operation modes, where EV1 provides power to EV2, but the power provided by EV1 is not enough to perform the battery charging of EV2. EV1 is discharging with a current of 20 A, while EV2 is charging with a current of 40 A. The figure shows the battery voltages (v_{bat1} and v_{bat2}) and currents (i_{bat1} and i_{bat2}), the currents drawn from the dc power grid, namely in the positive rail (i_{dcpos}), neutral rail (i_{dczer}) and negative rail (i_{dcneg}), and the voltages produced by the converters (v_{cv1} and v_{cv2}). Once again, i_{bat1} is negative, similar to V2G, but i_{bat2} is positive, similar to G2V. Regarding the dc power grid currents, it can be seen that i_{dcpos} is positive and i_{dcneg} is negative, meaning that the dc power grid is providing power. However, the average value of these currents is only 8 A, since the dc power grid only provides the power difference between the EV2 required power and the EV1 supplied power. Also, in this case, the currents i_{dcpos} and i_{dcneg} are symmetrical, with i_{dczer} presenting a null average value.

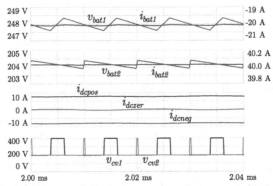

Fig. 8. Simulation results of the combination of V2V and G2V operation modes when EV1 is discharging with a current of 20 A and EV2 is charging with a current of 40 A.

Figure 9 shows the combination of V2V and V2G operation modes, where EV1 provides power to EV2, but the power provided by EV1 is more than the power required to perform the battery charging of EV2. EV1 is discharging with a current of 40 A, while EV2 is charging with a current of 20 A. The figure shows the battery voltages (v_{bat1} and v_{bat2}) and currents (i_{bat1} and i_{bat2}), the currents drawn from the dc power grid, namely in the positive rail (i_{dcpos}), neutral rail (i_{dczer}) and negative rail (i_{dcneg}), and the voltages produced by the converters (v_{cv1} and v_{cv2}). Once again, i_{bat1} is negative, similar to V2G, but i_{bat2} is positive, similar to G2V. Regarding the dc power grid currents, it can be seen that i_{dcpos} is negative and i_{dcneg} is positive, contrarily to the previous case, meaning that the dc power grid is receiving power. The average value of these currents is only 14.4 A, since the dc power grid only receives the power difference between the EV1 supplied power and the EV2 required power. As in the previous case, the currents i_{dcpos} and i_{dcneg} are symmetrical, with i_{dczer} presenting a null average value.

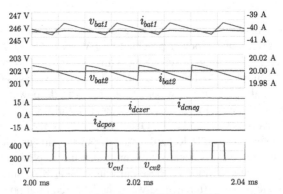

Fig. 9. Simulation results of the combination of V2V and V2G operation modes when EV1 is discharging with a current of 40 A and EV2 is charging with a current of 20 A.

In order to provide an overview of the obtained simulation results, Table 2 shows the average values of the main variables for each case, i.e., i_{bat1}, i_{bat2} and i_{dcpos}. The average values of i_{dcneg} and i_{dczer} are not presented since the average value of i_{dcneg} is always symmetrical with respect to i_{dcpos}, while the average value of i_{dczer} is always null, as expected and previously explained.

Table 2. Average value of the currents obtained in the simulation results.

CASE	I_{bat1}	I_{bat2}	I_{dcpos}
BALANCED G2V (Fig. 3)	20 A	20 A	23 A
UNBALANCED G2V (Fig. 4)	20 A	40 A	33.6 A
BALANCED V2G (Fig. 5)	−20 A	−20 A	−22.1 A
UNBALANCED V2G (Fig. 6)	−20 A	−40 A	−31.5 A
V2V (Fig. 7)	−16.3 A	20 A	0 A
V2V + G2V (Fig. 8)	−20 A	40 A	8 A
V2V + V2G (Fig. 9)	−40 A	20 A	−14.4 A

4 Conclusions

This paper presented a proposed electric vehicle battery charging station (EV-BCS) based on a bipolar dc power grid with vehicle-to-grid (V2G) and vehicle-to-vehicle (V2V) operation modes capability, besides the traditional battery charging operation mode (grid-to-vehicle – G2V). The presented EV-BCS is modular and uses three-level bidirectional dc-dc converters. A case scenario with two converters, and thus two EVs, was considered, aiming to validate all the operation modes (G2V, V2G and V2V, as well as the combination of V2V with G2V, and V2V with V2G). Moreover, in all operation modes it was considered an unbalanced operation from the EVs side, in order to emulate a real operation scenario and validate the proper operation of the EV-BCS. The obtained results were based on computational simulations and verify the correct operation of the EV-BCS in all cases, with balanced and unbalanced current consumption from the EVs, but with balanced currents from the bipolar dc power grid side.

Acknowledgments. This work has been supported by FCT – Fundação para a Ciência e Tecnologia within the Project Scope: UID/CEC/00319/2019. This work has been supported by the FCT Project DAIPESEV PTDC/EEI-EEE/30382/2017, and by FCT Project new-ERA4GRIDs PTDC/EEI EEE/30283/2017. Mr. Tiago J. C. Sousa is supported by the doctoral scholarship SFRH/BD/134353/2017 granted by the Portuguese FCT agency.

References

1. Chan, C.C., Wong, Y.S.: Electric vehicles charge forward. IEEE Power Energ. Mag. **2**(6), 24–33 (2004)
2. Milberg, J., Schlenker, A.: Plug into the future. IEEE Power Energ. Mag. **9**(1), 56–65 (2011)
3. Ansari, J., Gholami, A., Kazemi, A., Jamei, M.: Environmental/economic dispatch incorporating renewable energy sources and plug-in vehicles. IET Gener. Transm. Distrib. **8**(12), 2183–2198 (2014)
4. Knezovic, K., Martinenas, S., Andersen, P.B., Zecchino, A., Marinelli, M.: Enhancing the role of electric vehicles in the power grid: field validation of multiple ancillary services. IEEE Trans. Transp. Electrification **3**(1), 201–209 (2017)
5. Nguyen, H.N.T., Zhang, C., Zhang, J.: Dynamic demand control of electric vehicles to support power grid with high penetration level of renewable energy. IEEE Trans. Transp. Electrification **2**(1), 66–75 (2016)
6. Kempton, W., Tomić, J.: Vehicle-to-grid power implementation: from stabilizing the grid to supporting large-scale renewable energy. J. Power Sources **144**(1), 280–294 (2005)
7. Kesler, M., Kisacikoglu, M.C., Tolbert, L.M.: Vehicle-to-grid reactive power operation using plug-in electric vehicle bidirectional offboard charger. IEEE Trans. Industr. Electron. **61**(12), 6778–6784 (2014)
8. Monteiro, V., Pinto, J.G., Afonso, J.L.: Operation modes for the electric vehicle in smart grids and smart homes: present and proposed modes. IEEE Trans. Veh. Technol. **65**(3), 1007–1020 (2016)
9. Moghaddam, Z., Ahmad, I., Habibi, D., Phung, Q.V.: Smart charging strategy for electric vehicle charging stations. IEEE Trans. Transp. Electrification **4**(1), 76–88 (2018)
10. Yang, J., Wang, W., Ma, K., Yang, B.: Optimal Dispatching strategy for shared battery station of electric vehicle by divisional battery control. IEEE Access **7**, 38224–38235 (2019)
11. Zhang, Y., You, P., Cai, L.: Optimal charging scheduling by pricing for EV charging station with dual charging modes. IEEE Trans. Intell. Transp. Syst. **20**(9), 3386–3396 (2019)
12. Morstyn, T., Crozier, C., Deakin, M., McCulloch, M.D.: Conic optimization for electric vehicle station smart charging with battery voltage constraints. IEEE Trans. Transp. Electrification **6**(2), 478–487 (2020)
13. Dai, Q., Cai, T., Duan, S., Zhao, F.: Stochastic modeling and forecasting of load demand for electric bus battery-swap station. IEEE Trans. Power Delivery **29**(4), 1909–1917 (2014)
14. Tan, X., Qu, G., Sun, B., Li, N., Tsang, D.H.K.: Optimal scheduling of EV-BCS serving electric vehicles based on battery swapping. IEEE Trans. Smart Grid **10**(2), 1372–1384 (2019)
15. Liu, X., Zhao, T., Yao, S., Soh, C.B., Wang, P.: Distributed operation management of battery swapping-charging systems. IEEE Trans. Smart Grid **10**(5), 5320–5333 (2019)
16. Ahmad, F., Saad Alam, M., Saad Alsaidan, I., Shariff, S.M.: Battery swapping station for electric vehicles: opportunities and challenges. IET Smart Grid **3**(3), 280--286 (2020)
17. Vasiladiotis, M., Rufer, A.: A modular multiport power electronic transformer with integrated split battery energy storage for versatile ultrafast EV charging stations. IEEE Trans. Industr. Electron. **62**(5), 3213–3222 (2015)
18. Yan, Q., Zhang, B., Kezunovic, M.: Optimized operational cost reduction for an EV charging station integrated with battery energy storage and PV generation. IEEE Trans. Smart Grid **10**(2), 2096–2106 (2019)
19. Deng, Y., Zhang, Y., Luo, F.: Operational planning of centralized charging stations using second-life battery energy storage systems. IEEE Trans. Sustain. Energy **3029**(c), 1 (2020)
20. Patterson, B.T.: DC, come home: DC microgrids and the birth of the 'enernet.' IEEE Power Energ. Mag. **10**(6), 60–69 (2012)

21. Kumar, D., Zare, F., Ghosh, A.: DC microgrid technology: system architectures, AC grid interfaces, grounding schemes, power quality, communication networks, applications, and standardizations aspects. IEEE Access **5**, 12230–12256 (2017)
22. Sun, L., Zhuo, F., Wang, F., Zhu, T.: A nonisolated bidirectional soft-switching power-unit-based DC–DC converter with unipolar and bipolar structure for DC networks interconnection. IEEE Trans. Ind. Appl. **54**(3), 2677–2689 (2018)
23. Guo, L., Yao, G., Huang, C., Zhou, L.: Bipolar output direct-coupled DC–DC converter applied to DC grids. J. Eng. **2019**(16), 1474–1479 (2019)
24. Perera, C., Salmon, J., Kish, G.J.: Multiport converter with independent control of AC and DC power flows for bipolar DC distribution. IEEE Trans. Power Electron. **8993**(c), 1 (2020)
25. Rivera, S., Wu, B., Kouro, S., Yaramasu, V., Wang, J.: Electric vehicle charging station using a neutral point clamped converter with bipolar DC bus. IEEE Trans. Industr. Electron. **62**(4), 1999–2009 (2015)
26. Tan, L., Wu, B., Yaramasu, V., Rivera, S., Guo, X.: Effective voltage balance control for bipolar-DC-bus-fed EV charging station with three-level DC–DC fast charger. IEEE Trans. Industr. Electron. **63**(7), 4031–4041 (2016)
27. Tan, L., Wu, B., Rivera, S., Yaramasu, V.: Comprehensive DC power balance management in high-power three-level DC–DC converter for electric vehicle fast charging. IEEE Trans. Power Electron. **31**(1), 89–100 (2016)
28. Rivera, S., Wu, B.: Electric vehicle charging station with an energy storage stage for split-DC bus voltage balancing. IEEE Trans. Power Electron. **32**(3), 2376–2386 (2017)
29. Kim, S., Cha, H., Kim, H.-G.: High-efficiency voltage balancer having DC–DC converter function for EV charging station. IEEE J. Emerg. Sel. Topics Power Electron. **6777**(c), 1 (2019)
30. Sousa, T.J.C., Monteiro, V., Fernandes, J.C.A., Couto, C., Melendez, A.A.N., Afonso, J.L.: New perspectives for vehicle-to-vehicle (V2V) power transfer. In: IECON 2018- 44th Annual Conference of the IEEE Industrial Electronics Society, pp. 5183–5188 (2018)
31. Taghizadeh, S., Jamborsalamati, P., Hossain, M.J., Lu, J.: Design and Implementation of an advanced vehicle-to-vehicle (V2V) power transfer operation using communications. In: 2018 IEEE International Conference on Environment and Electrical Engineering and 2018 IEEE Industrial and Commercial Power Systems Europe (EEEIC / I&CPS Europe), pp. 1–6 (2018)
32. Mou, X., Zhao, R., Gladwin, D.T.: Vehicle to vehicle charging (V2V) bases on wireless power transfer technology. In: IECON 2018 - 44th Annual Conference of the IEEE Industrial Electronics Society, pp. 4862–4867 (2018)
33. Mou, X.: "Vehicle-to-Vehicle charging system fundamental and design comparison. IEEE Int. Conf. Ind. Technol. (ICIT) **2019**, 1628–1633 (2019)
34. Shen, M., Peng, F.Z., Tolbert, L.M.: Multilevel DC–DC power conversion system with multiple DC sources. IEEE Trans. Power Electron. **23**(1), 420–426 (2008)
35. Yilmaz, M., Krein, P.T.: Review of battery charger topologies, charging power levels, and infrastructure for plug-in electric and hybrid vehicles. IEEE Trans. Power Electron. **28**(5), 2151–2169 (2013)
36. Monteiro, V., Sousa, T.J.C., Sepúlveda, M.J., Couto, C., Lima, A., Afonso, J.L.: A proposed bidirectional three level DC-DC power converter for applications in smart grids: an experimental validation. In: 2019 IEEE SEST International Conference on Smart Energy Systems and Technologies, Porto, Portugal, September 2019

Smart Charging and Renewable Grid Integration - A Case Study Based on Real-Data of the Island of Porto Santo

Leo Strobel[1], Jonas Schlund[1], Veronika Brandmeier[2], Michael Schreiber[2], and Marco Pruckner[1(✉)]

[1] Energy Informatics, Friedrich-Alexander-University Erlangen-Nürnberg, Erlangen, Germany
{leo.strobel,jonas.schlund,marco.pruckner}@fau.de
[2] The Mobility House, Munich, Germany
{veronika.brandmeier,michael.schreiber}@mobilityhouse.com

Abstract. The penetration of battery electric vehicles is increasing. Due to their ability to store electrical power and to shift charging events, they offer a wide range of opportunities with regard to renewable grid integration and lowering the overall CO2 emissions. This is particularly evident for isolated microgrids, such as the Portuguese island of Porto Santo. In this paper, we conduct a data analysis of real-world charging data of 20 electric vehicles operated on the island of Porto Santo. We provide insights into the charging behavior of different users and analyze the opportunity of smart charging for better renewable grid integration using linear optimization models. The data analysis shows that drivers prefer home rather than publicly available charging stations, flexible charging events occur mostly overnight and the charging flexibility of the fleet decreases over the project duration. With regard to make charging more flexible, we can see that smart charging can help to raise the share of electricity generated by renewable energy sources for charging the electric vehicles to up to 33%.

Keywords: Electric vehicles · Vehicle-2-Grid · Grid integration · Smart charging

1 Introduction

All over the world the penetration of electric vehicles (EVs) is increasing. According to the Global EV Outlook [8], the global EV fleet reached a number of 5.1 million units in 2018. An increasing number of EVs comes with challenges and opportunities. Distribution system operators (DSOs) face new challenges in operating their networks, especially in the case of uncoordinated charging during the evening peak hours. On the other hand, smart EV charging strategies can contribute to stabilize distribution grids by supplying energy storage for the integration of volatile renewable energy sources (RES), such as solar or wind

© ICST Institute for Computer Sciences, Social Informatics and Telecommunications Engineering 2021
Published by Springer Nature Switzerland AG 2021. All Rights Reserved
J. L. Afonso et al. (Eds.): SESC 2020, LNICST 375, pp. 200–215, 2021.
https://doi.org/10.1007/978-3-030-73585-2_13

energy [15]. Additionally, bi-directional charging and Vehicle-to-Grid technologies (V2G) can enable even more benefits in terms of renewable grid integration and grid stabilization.

According to a recently published report [5], there are more than 60 projects involving thousands of EVs and chargers. One of the projects mentioned is the plan to make Portuguese island Porto Santo completely fossil-free by targeting multiple action areas. Within the area of sustainable mobility, a joint project was launched by Groupe Renault and Empresa de Electricidade da Madeira (EEM) with the support and technology of The Mobility House to create a smart electric ecosystem on the island of Porto Santo. Thereby, smart charging of EVs as well as V2G and stationary battery storage systems play an important role. In order to provide better insights on the EV charging behavior, we contribute a comprehensive data analysis of real-world charging data of 20 EVs for a time horizon of six months. Thereby, we also discuss the opportunity of smart charging in the context of vehicle grid integration (VGI). Based on this information, we investigate smart charging potentials to better integrate electricity generated by RES on the island of Porto Santo.

The remainder of this paper is structured as follows: In Sect. 2 we give an overview of related work in the field of VGI on islands as well as data analysis of EV charging data. More information of the Porto Santo project is provided in Sect. 3. Section 4 gives insights into the charging processes and the user behavior. In Sect. 5 we formulate an optimization problem for the maximization of RES usage under consideration of the charging flexibility and user requirements and present respective results. Finally, Sect. 6 concludes the paper with a short summary and an outlook on future work.

2 Related Work

2.1 Vehicle Grid Integration on Islands

The topic of VGI on islands has already been analyzed from different research perspectives. A good overview on island applications of EVs is provided by Gay et al. [7]. The authors review several studies that explore the effect of EV grid integration in terms of V2G services and greenhouse gas emissions on isolated island grids. Additionally, they present a case study for the Caribbean island of Barbados in order to link the principles of V2G services to an existing small island developing state. One of the main findings of the case study addresses the concerns about the impact of EVs on the isolated electricity grids including overloaded distribution feeders and transformers. Possible solutions include smart charging approaches and V2G services where EVs can become a key grid asset. In Pina et al. [12] and Verzijlbergh et al. [16] the potential of EVs on the Portuguese Island of Flores is investigated. Pina et al. [12] present a scenario-based study on the impact of EVs on an isolated island energy system. Under different EV penetration rates and different recharging strategies they investigate the primary energy consumption and CO_2 emissions. The work of Verzijlbergh et al. [16] is focused on the investigation of EVs to support the grid integration of a high share of renewable energy sources on the island of Flores. The authors show that there is a large potential for saving CO_2 emissions compared to diesel generation units and diesel-fueled vehicles. Binding et al. [1] present a platform to optimally integrate EVs on the Danish island of Bornholm. The platform is used to realize the potential of using V2G services in a virtual power plant.

The literature survey reveals that some studies already addressed the grid integration of EVs on islands. However, all approaches have potential for improvement either with regard to the scope or the use of real driving behavior patterns. For instance, Binding et al. [1] uses an agent-based simulation environment to model three different types of EVs (commuter cars, taxis, and family cars); the charging behavior of the EVs is also simulated. In the case of the island of Flores, Verzijlbergh et al. [16] and Pina et al. [12] use a survey of driving patterns to model the driving behavior and charging needs of EVs. In contrast to previous work, this paper present real-word data from a fleet of 20 EVs including plug-in time, plug-out time, charged energy, charging station identification as well as anonymized driver and vehicle identification for the small island of Porto Santo.

2.2   Data Analysis of EV Charging Data

Although EV charging data is sparse, there are some recently published studies about real EV charging or driving behavior patterns available. Lee et al. [9] provide a dynamic dataset of EV charging which includes over 30,000 charging sessions collected from two workplace charging sites in California. The ElaadNL dataset [3,6,13] contains 400,000 events on 1,750 public accessible charging stations distributed over the entire Netherlands. According to Flammini et al. [6] the dataset shows important key figures, such as charge time, idle time, connected time, power, and energy. Chen et al. [2] analyze charging characteristics such as charging time, charging duration and charged capacity on real data from Nanjing, China and derive probability distributions and correlations of these features. Xydas et al. [17] develop a characterization framework for the EV charging demand based on very detailed charging data from the UK. Thereby, the authors develop a data mining model to analyze the characteristics of EV charging demand in a geographical area.

As our literature review shows, there are a few studies available which demonstrate the necessity of real-world data in the EV charging context. We make an important contribution to these works by analyzing the charging data in the Porto Santo project. Since the EV charging dataset of Porto Santo is unique in containing specific, anonymized, driving and vehicle consumption data as well as home and public charging, a more precise analysis of the user behavior is possible. Additionally, the EV charging on Porto Santo has already moved on from uncoordinated charging to enable EV drivers to freely set departure time and state of charge (SOC) that needs to be guaranteed. Therefore, the willingness of EV drivers to charge flexibly can be exploited. The fact that Porto Santo is an island leads to advantages for the data analysis. The grid of Porto Santo is not connected to the outside world, therefore the grid impact and RES share of EV charging can be determined without the difficulties of accounting for the import and export of electricity.

3 Project Porto Santo

3.1 General Information

Groupe Renault and Empresa de Electricidade da Madeira (EEM), the Madeira electricity company, with the support and technology of The Mobility House jointly launched a project to create a smart electric ecosystem on the island of Porto Santo. The aim of the project is to support Porto Santo on its transition of becoming the world's first CO_2-free island, to ensure a reliable and intelligent power grid and to avoid cost-intensive grid expansion. The project supports reaching these goals through four action areas: EVs, stationary energy storage, smart charging and V2G charging. The EVs and stationary batteries are used to balance renewable production volatility, which is facilitated by an intelligent smart charging controller and technology.

Porto Santo is a Portuguese island located 43 km northeast of Madeira Island. The island's main economic area, tourism, is characterized by a high seasonality of economic, social and cultural activities which focus on about three months per year. During high season, the population of the 5,000 inhabitants is increased to up to 20,000 by part-time inhabitants and tourists. During these periods, the energy demand is particularly high.

The grid on Porto Santo is a closed system and not connected to Madeira Island. Currently, the electricity demand of 33 GWh (2017) is mainly covered by diesel generators, presenting approximately 85% of the production. The remaining 15% are generated by a solar park of 2 MW and various smaller PV plants, summing up to 0.43 MW as well as a wind turbine of 0.66 MW [4]. The potential for expansion of renewable generation, particularly of photovoltaics, exceeds the energy demand of the island by far.

In a first phase, 20 Renault brand electric vehicles are handed over to public institutions (e.g., police), private companies (such as taxi drivers) and private individuals, who use them for their everyday mobility needs. Out of the 20 EVs, fourteen are Renault ZOE with a battery capacity of 41 kWh and six are Kangoo Z.E with 33 kWh. To make this experience possible for many users, about half of all vehicles change their owners every two months. EV drivers have the possibility to charge their EVs at home (accessible only for one user) and at public (accessible for all project participants) charging stations (total of 33 charging stations). All charging stations are equipped with a controller from The Mobility House and an internet connection. This enables smart charging and monitoring of the charging process. The charging stations are single-phased and have a rated power or 7.4 kW. The uni-directional EV fleet is complemented by two bi-directional Renault EVs with 41 kWh battery capacity and 11 kW charging and discharging power. In terms of costs, home and public charging was free of charge for all users during the considered time horizon.

3.2   Data Sets

The collected, anonymized data contains records of the charging stations' utilization on different levels of granularity. Overall, data for 33 charging stations is available. 8 are publicly available and 25 (including the two V2G enabled stations) are located at private locations.

One part of the data is an aggregated dataset of the entire EVs fleet on Porto Santo. It is available for six months from February 2019 to July 2019 in a timely resolution of 10 s. Every record of this dataset contains the current charging capability, the charged amount of energy as well as the *must* charge value. The charging capability reflects the total possible charging power of all plugged-in EVs in a certain time step.

Smart charging on Porto Santo works under the concept of scheduling. For every charging event exists a schedule determining a minimal SOC, that has to be reached at a specified time. The charging itself is dispatched via a logic preferring renewable energy. However, if this strategy is not able to fulfill the schedule the EV switches into a *must*-charge state, in which the EV immediately charges. Since February 2019 the EV drivers can freely set departure time and goal SOCs via an app. If they do not, the charging station will automatically fall back on a default schedule, which guarantees a SOC of 80% at 7am and 5 or 7pm. As the possibility of monitoring the SOC is only available since August 2019, the schedules always assume that EVs have a SOC of 30% when they plug in at the charging station and dispatch accordingly.

Another dataset contains detailed data records of individual charging events of all EVs for about three months between May, 16th 2019 and August, 6th 2019. This data set contains about 700 charging events each associated with a specific driver id, vehicle id, charging station id, plug-in time and plug-out time as well as the amount of charged energy. This dataset is used i.a. for the determination of individual charging flexibility. In order to be able to analyze the share of electricity generated by RES which is used for charging EVs, Porto Santo's electrical energy generation between January 2018 and August 2019 is also provided.

4   Data Analysis

The data analysis focuses on user behavior and the flexibility it provides in terms of shifting energy consumption towards time periods with a higher share of RES.

4.1   Charging Demand and Electricity Generation

Our analysis of the individual charging events yields that EVs on Porto Santo are often charged at home. Figure 1 shows a boxplot of the number of EVs that are plugged-in at any given hour of the day at home charging stations. EVs are mostly plugged in at home during non-working hours between 11pm and 6am with three vehicles on average at night. The plugged-in vehicles at

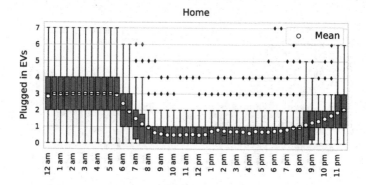

Fig. 1. Plugged-in EVs over clock time at home charging stations in the individual data.

public charging stations do not significantly vary during the day - especially not during working hours. This is contrary to patterns seen in Flanders, Belgium [15]. Reasons are shorter driving distances than in Belgium and that not all public charging stations are installed at sites where people work. In general, public charging events show no significant peaks over the course of a day. We also analyzed differences between weekdays and weekends for both home and public charging events. Both follow the average course apart from the difference that a slightly lower amount of plugged-in vehicles are observable on weekends. On public charging stations the average per day drops from 0.44 plugged-in vehicles on weekdays to 0.31 on weekends and on home charging stations it drops from 1.58 to 1.25, respectively.

In Fig. 2 the capability of the aggregated charging profiles follows the same course as the plugged-in vehicles. This is not surprising since most charging stations have the same rated power of 7.4 kW. Due to scheduling of the charging processes the demand does not follow the expected curve of dumb charging. As shown in Fig. 2, the main demand is between 3 and 7am. The *must* charge course gives insight about the users choice of schedule. Higher amounts of *must* charge indicate that many scheduled departure times are imminent. Judging from Fig. 2 we observe that *must* charge is always high shortly before a default departure time (7am, 5pm and 7pm) is reached. This suggests that many drivers either stuck with the default schedule or chose a very similar one.

The EV charging process shows significant deviations from day to day. The 30-min resoluted capability has a standard deviation of 13.6 kW and can go up to 73 kW. *Must* charge has a standard deviation of 6 kW and demand of 7 kW, both show a maximum of 65 kW. The charging demand is especially noisy in the time frames from 4 to 7am and from 2 to 6pm, with an average standard deviation of 13 kW and 5 kW respectively. During the entire remaining day it remains below 2.5 kW. This suggest that drivers are very likely to charge during these time frames but not necessarily every day. On average every vehicle is only charged every third day.

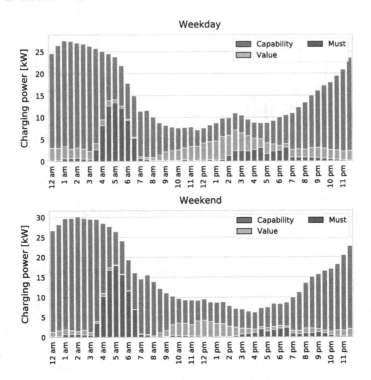

Fig. 2. Charging power over clock time at weekdays and weekends in the aggregated data (overlapping bars).

Weekends have an overall lower charging energy demand. On average the daily energy demand is 100 kWh, while it is only 87 kWh on weekends. Figure 2 shows that the overall course of charging differs mostly in the morning peak from 3 to 7am and in the afternoon 2 to 5pm. The morning peak is considerably higher while the mostly freely charged afternoon peak is almost non existent. This would suggest that at weekends even less charging events occur during the day, however this contradicts the findings from the individual data, where plug in number decreased uniformly over the entire weekend.

Figure 3 shows the average electricity generation over clock time. The average share of RES in the investigated time period between January and June 2019 is 15.1%. In the winter the wind and PV generation are almost equal. In the summer PV generates about four times as much electricity than wind. Thus, over the course of a year most of the renewable energy is generated by PV and the amount of total RES generation is very low during nighttime. As the charging strategy aims at maximizing the RES share, this leads to delayed charging of all users that arrive after sunset. However, since most users stuck with the default schedule demanding 80% SOC at 7am, the charging stations have to switch into *must* charge mode at around 4am to fulfill the schedule. This explains the high demand during 3 and 7am in Fig. 2.

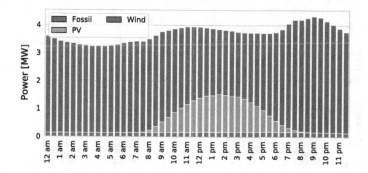

Fig. 3. Electricity generation over clock time for spring 2019 (stacked bars).

Most users stuck with their default schedule. Therefore overnight charging events are very flexible. However, it is problematic that the share of renewables is lowest exactly in this time frame. The desynchronized behavior of the charging capability and the renewable energy generation is a major obstacle in achieving a high share of RES for the EV charging on Porto Santo. Thus, the current charging strategy in Porto Santo has a substantial impact by increasing the RES proportion of charging to 16.4% even though the favored overnight plug-in times do not match the PV production.

4.2   User Behavior and Charging Flexibility

A measure for the flexibility of charging events (or any other flexible load) is the *time flexibility* [10,11,14]. The time flexibility ϕ is defined as the maximum amount of time the charging energy can be shifted. It is calculated according to Eq. 1 by the difference between the total plugged-in duration $\Delta t_{\text{plugged-in}}$ in h and the time it would take to charge the required energy amount E_{req} at the maximum power rate P_{max}.

$$\phi = \Delta t_{\text{plugged-in}} - \frac{E_{\text{req}}}{P_{\text{max}}} \tag{1}$$

Figure 4 shows the distribution of the user behavior metrics of individual charging events: a) charged energy, b) plugged-in duration and c) the time flexibility. Figure 4c) shows that most public charging events have less than three hours of time flexibility. In general, the public charging events show significantly less time flexibility than those at homes. At home the available time flexibility differs greatly between the different cars. While the group of ZOEs has a higher median of 4 h, the Kangoos have one peak at 9 h of time flexibility and another peak at 1 h. Similar distributions can be seen for the plugged-in durations. The mean charging energy is significantly higher for the ZOEs, which suggests a correlation with the battery size. Figure 4d) shows that charging events at home are more numerous than in public. Interestingly many drivers only frequent either

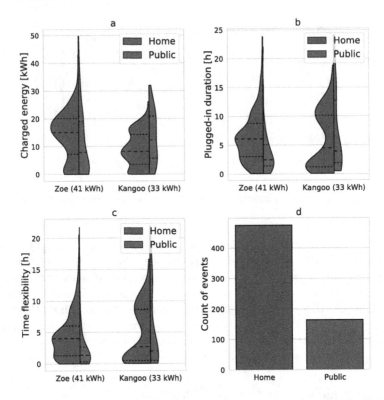

Fig. 4. Distribution of metrics for charging events in the individual data. Divided up by station and vehicle kind. The dashed line with longer dashes is the median, the lines with shorter dashes represent quartiles.

public or home stations. Out of the 24 in the data set only 7 frequented both types of charging stations.

The different users show very individual behavior. In Fig. 5 plug-in and plug-out times of all the events of the individual charging events data set are displayed in a bubble plot. The size of the bubbles reflects the charged energy and some users are highlighted by colors. The cloud on the lower right are the overnight charging events whilst the cloud in center stretching to the top right consists of intraday events. The composition of these two clouds differs. Public charging events make up for 25.5% of the intraday events and only for 11.5% of the overnight events.

Some users, like user 4 shown in blue, behave according to a consistent pattern, while many others, like user 8 shown in green, are hard to predict. Nonetheless some overall patterns become clear. Firstly, the duration of intraday events is shorter than five hours in 84% of the cases. The duration of overnight charging events differs greatly, however they almost always plug out between 6 and 9am. Secondly, in the intraday case charging energy increases very distinctly with longer plugged-in durations, while the charging energy is more uniformly

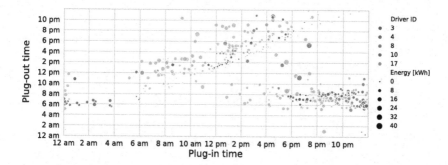

Fig. 5. Scatter plot of charging events over start and stop clock time. Some driver IDs are highlighted.

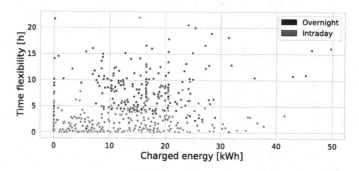

Fig. 6. Scatter plot of time flexibility over charged energy per charging event.

distributed in the overnight case. This is confirmed by the Bravais-Pearson correlation coefficient between the charging energy and plugged-in duration. For intraday events it is 0.6 while it is 0.37 overnight. This suggests that most intraday events are terminated soon after the energy demand has been supplied.

Summarizing, the primary time frame for flexible charging is overnight. A plot of the time flexibility over the charged energy in Fig. 6 shows this very clearly. This observation strengthens the challenge mentioned in Sect. 4.1, that flexible charging options are desynchronized with RES generation.

Another challenge is the overall development of the approximated time flexibility $\hat{\phi}$ of the entire EV fleet during the pilot project since the introduction of free schedule choice. This flexibility is approximated with Eq. 2, on the basis of Eq. 1, for weekly intervals. The plugged-in duration is substituted by the duration of a week Δt_{week} and the maximum power rate is substituted by the average capability of the EV fleet $\bar{P}_{\text{cap}}$ (see Sect. 3.2). In this context the required energy E_{req} is the total demand of the EV fleet during the week.

$$\hat{\phi} = \Delta t_{\text{week}} - \frac{E_{\text{req}}}{\bar{P}_{\text{cap}}} \tag{2}$$

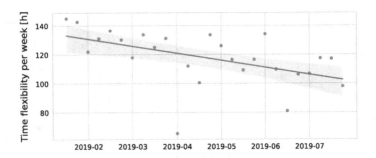

Fig. 7. Flexibility of charging since the implementations of individual schedules. Estimated from the aggregated data in weekly intervals.

Table 1. Variable definition overview

Symbol	Description
t	Model time
T	Length of the optimization horizon
Δt	Time step length
P_t^{RES}	Electricity generated by RES at time step t
P_t^{total}	Total electricity generation at time step t
P_t^{charge}	Total charging power at time step t
E^{real}	Total charging demand
P_t^{max}	Maximum charging power rate

As shown in Fig. 7, the trend of $\hat{\phi}$ is downwards. The graph shows weekly decline in flexibility since February 2019 (beginning of the evaluation). This means that people tend to plug in less with ongoing project time, which arises the question of suitable incentives to foster plugging-in the vehicle whenever possible to increase the available charging flexibility.

5   Optimization

To analyze the optimal integration of RES into the EV sector of Porto Santo three different optimizations have been carried out. All basic definitions are summarized in Table 1.

5.1   Methodology

All three optimizations share the same mathematical structure which is given in Eq. 3a to Eq. 3c. They only differ in the input dataset and the optimization horizon. The objective is to maximize the RES share of the charged energy. Equation 3a shows the objective function. Additionally the maximum power rate (P_t^{max}) at every time step must not be exceeded (Eq. 3c), discharging is not allowed and the energy demand (E^{real}) in a given time period T must be fulfilled (Eq. 3b). The time step length Δt is always 15 min.

$$\underset{P^{\text{charge}}}{\text{maximize}} \quad E^{\text{real}^{-1}} \cdot \sum_{t \in T} (\frac{P_t^{\text{RES}}}{P_t^{\text{total}}} \cdot P_t^{\text{charge}} \cdot \Delta t) \tag{3a}$$

$$\text{subject to} \quad \sum_{t \in T} (P_t^{\text{charge}} \cdot \Delta t) = E^{\text{real}}, \tag{3b}$$

$$0 \leq P_t^{\text{charge}} \leq P_t^{\text{max}}, \qquad \forall t \in T \tag{3c}$$

Optimization (Opt.) 1 and 2 are calculated on the aggregated dataset. In this context P_t^{max} describes the capability and E^{real} the sum of the energy demand of the entire EV fleet in T. The Optimization is not carried out over the entire data set at once, but in separate time segments of fixed length. Opt. 1 and 2 only differ in their choice of segmentation. Opt. 1 is carried out in day long segments ($T = 24h$) allowing energy shift between night and day. As discussed in Sect. 4 the two most distinct time frames of charging are intraday and overnight. This is represented in Opt. 2, in which the periods are divided into two groups: intraday for the time frame between 7am to 7pm and overnight from 7pm to 7am (with $T = 12h$).

Both Opt. 1 and 2 only take the behavior and demand of the entire EV fleet into consideration neglecting individual constraints. To determine the impact off this approximation a third approach (Opt. 3) is implemented with the data set of the individual charging sessions. This data set is already segmented by event. Thus in Opt. 3, the time period T is the time frame between plug-in and plug-out time, E^{real} is the actual charged energy of that event and P_t^{max} refers to the maximum charging rate of the station and is therefore constant.

The results of Opt. 2 show an idealization of the current user behavior thus an overestimation of the energy constraint. The energy of charging events can be shifted between 7pm to 7am and between 7am and 7pm. Especially the intraday events on Porto Santo have a duration shorter than 12 h. The results of Opt. 1 would require a significant change in the user behavior. The EVs would need to be plugged-in both in public and at home. Opt. 3 shows the best integration possible with the current behavior.

5.2 Results

The share of RES during the evaluation period was 15.1%. In this first project phase, the fleet was charged with 16.4% without considering forecasts. The slightly higher share is already a significant improvement compared to a scenario where EVs are charged immediately after plug-in, since the high share of overnight charging correlates with a low RES share on Porto Santo. However, Opt. 1 shows a high increase in the RES integration. In total, the share of RES can be increased from 16.4% in the real case to 26.5%. Opt. 2 reaches 20.1%.

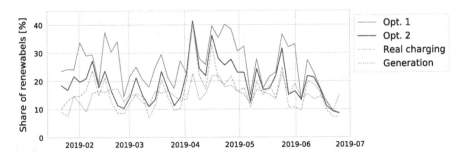

Fig. 8. Share of RES for two optimal cases, the real case and the generation over the project duration. Averaged over 3 and a half days.

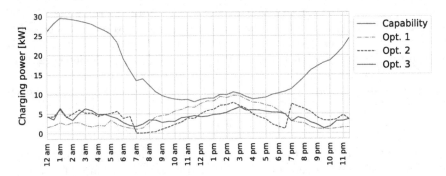

Fig. 9. Charging power over clock time for each optimization case.

Figure 8 shows the renewable share of the charged energy over the project duration.

The current charging strategy mostly follows this baseline with short but significant advantages in certain time frames, like in the first quarter of April. Perfect dispatch with separation of night and day charging, represented by Opt. 2, would increase the size of these positive windows.

The biggest opportunity for higher RES integration lies in removing the barrier between intraday and overnight charging. Figure 8 shows that Opt. 1 leads to an overall strong improvement compared to the real charging case. Figure 9 shows the daily average dispatch of all scenarios. It becomes clear that the main advantage of Opt. 1 lies in the higher integration of PV, since most charging gets shifted to daytime. As PV generation is also highly predictable on a daily basis, this advantage should therefore be transferable to the real world.

In order to realize this, strong incentives to plug-in the vehicle at day and at night should be established. However, Opt. 1 and 2 overestimate the results as they neglect the constraints from individual users. The optimization based on the real user behavior, Opt. 3, is an upper bound for the effectiveness of smart charging strategies that do not impact user comfort. Due to missing data, it is only calculated for a shorter time frame. However, it shows a very similar RES

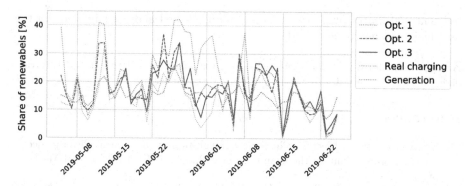

Fig. 10. Share of RES for each optimal cases, the real case and average in generation over the Project duration. Average over single days.

integration as Opt. 2 (see Fig. 10). Opt. 2 can predict the more precise Opt. 3 with a coefficient of determination (R^2) of 0.75. The share of RES from Opt. 3 is at 18.3%. For comparison in the same time frame Opt 2. has a RES share of 18.2%. This shows that the methodology of Opt. 2 is able to estimate Opt. 3 well. The fact that Opt. 3 and Opt. 2 have almost the same results shows that for the current scenario on Porto Santo, smart charging strategies are limited to an increase of the RES share of about 5% points (or 33%) compared to the RES proportion in the power grid. As long as the driver behavior does not change, e.g., by means of the introduction of variable charging costs or new incentice schemes like gamification or financial remuneration, a full usage of the high PV peaks will remain limited.

6   Conclusion

This research work addressed the introduction of EVs on the Portuguese island of Porto Santo that aims to become the world's first CO_2-free island. To support the achievement of the objectives, Groupe Renault and Empresa de Electricidade da Madeira (EEM) with the support and technology of The Mobility House, jointly launched a project to create a smart electric ecosystem on Porto Santo. In this work, we gained interesting insights of the charging behavior of different user groups on Porto Santo driving 20 EVs for a six-month period. Thereby, we also studied the opportunity of flexible charging by introducing three linear optimization models. Results showed, that smart charging strategies on Porto Santo can help to increase the RES share to 33% compared to the RES proportion in the grid.

As we proceed, we intend to include more real-world data after the Porto Santo project is finished. In parallel, we will start working on a holistic model for the energy system of Porto Santo in order to conduct a scenario-based study for Porto Santo's transition towards the first CO2-free island. For the operations on Porto Santo, forecasts and optimization should be added to the smart charging logic in a second project phase. Furthermore, incentive schemes for more frequent and longer plug-in events may be developed and their influece on the user behavior may be analyzed.

Acknowledgment. The authors would like to thank Groupe Renault and Electricidade da Madeira for data provision and useful discussions.

References

1. Binding, C., et al.: Electric vehicle fleet integration in the danish edison project - a virtual power plant on the island of bornholm. In: IEEE PES General Meeting, pp. 1–8 (2010)
2. Chen, Z., Zhang, Z., Zhao, J., Wu, B., Huang, X.: An analysis of the charging characteristics of electric vehicles based on measured data and its application. IEEE Access **6**, 24475–24487 (2018)
3. Develder, C., Sadeghianpourhamami, N., Strobbe, M., Refa, N.: Quantifying flexibility in EV charging as DR potential: analysis of two real-world data sets. In: 2016 IEEE International Conference on Smart Grid Communications (SmartGridComm), pp. 600–605 (2016)
4. Empresa de Electricidade da Madeira (EEM): Internal report (2018)
5. Everoze Partners Ltd. and EVConsult: V2G Global Roadtrip: Around the World in 50 Projects - Lessons learned from fifty international vehicle-to-grid projects (2019)
6. Flammini, M.G., Prettico, G., Julea, A., Fulli, G., Mazza, A., Chicco, G.: Statistical characterisation of the real transaction data gathered from electric vehicle charging stations. Electric Power Syst. Res. **166**, 136–150 (2019)
7. Gay, D., Rogers, T., Shirley, R.: Small island developing states and their suitability for electric vehicles and vehicle-to-grid services. Utilities Policy **55**, 69–78 (2018)
8. International Energy Agency: Global ev outlook 2019 - scaling-up the transition to electric mobility (2019)
9. Lee, Z.J., Li, T., Low, S.H.: ACN-data: analysis and applications of an open EV charging dataset. In: Proceedings of the Tenth ACM International Conference on Future Energy Systems, pp. 139–149. e-Energy 2019, Association for Computing Machinery, New York (2019)
10. Nakahira, Y., Chen, N., Chen, L., Low, S.: Smoothed least-laxity-first algorithm for ev charging. In: Proceedings of the Eighth International Conference on Future Energy Systems. p. 242–251. e-Energy 2017, Association for Computing Machinery, New York, NY, USA (2017)
11. Neupane, B., Šikšnys, L., Pedersen, T.B.: Generation and evaluation of flex-offers from flexible electrical devices. In: Proceedings of the Eighth ACM International Conference on Future Energy Systems, e-Energy 2017, pp. 143–156. ACM, New York (2017)
12. Pina, A., Baptista, P., Silva, C., Ferrão, P.: Energy reduction potential from the shift to electric vehicles: the flores island case study. Energy Policy **67**, 37–47 (2014)
13. Sadeghianpourhamami, N., Refa, N., Strobbe, M., Develder, C.: Quantitive analysis of electric vehicle flexibility: a data-driven approach. Int. J. Electr. Power Energy Syst. **95**, 451–462 (2018)
14. Schlund, J., Pruckner, M., German, R.: FlexAbility - modeling and maximizing the bidirectional flexibility availability of unidirectional charging of large pools of electric vehicles. In: Proceedings of the Eleventh ACM International Conference on Future Energy Systems. e-Energy 2020. ACM, New York (2020)

15. Van Roy, J., Leemput, N., Geth, F., Büscher, J., Salenbien, R., Driesen, J.: Electric vehicle charging in an office building microgrid with distributed energy resources. IEEE Trans. Sustainable Energy **5**(4), 1389–1396 (2014)
16. Verzijlbergh, R.A., Ilić, M.D., Lukszo, Z.: The role of electric vehicles on a green island. In: 2011 North American Power Symposium, pp. 1–7 (2011)
17. Xydas, E., Marmaras, C., Cipcigan, L.M., Jenkins, N., Carroll, S., Barker, M.: A data-driven approach for characterising the charging demand of electric vehicles: a UK case study. Appl. Energy **162**, 763–771 (2016)

Author Index

Printed in the United States
by Baker & Taylor Publisher Services

Printed in the United States
by Baker & Taylor Publisher Services